Air Launch!

A Pictorial History of Airborne Weapons

Bill Holder
&
Scott Vadnais

Schiffer Military/Aviation History
Atglen, PA

Acknolwedgments

I. Wolfram Lauther, Daimler-Benz Aerospace
2. Don Bernstein, Raytheon Company
3. Claude Morris, Arnold Engineering Development Center
4. David Menard, Research Department, Air Force Museum
5. Laura Romesberg, ASC History Office, WPAFB

Book Design by Ian Robertson.

Library of Congress Control Number: 2001090707

Printed in China.
ISBN: 0-7643-1392-4

We are interested in hearing from authors with book ideas on related topics.

Published by Schiffer Publishing Ltd.
4880 Lower Valley Road
Atglen, PA 19310
Phone: (610) 593-1777
FAX: (610) 593-2002
E-mail: Schifferbk@aol.com.
Visit our web site at: www.schifferbooks.com
Please write for a free catalog.
This book may be purchased from the publisher.
Please include $3.95 postage.
Try your bookstore first.

In Europe, Schiffer books are distributed by:
Bushwood Books
6 Marksbury Avenue
Kew Gardens
Surrey TW9 4JF
England
Phone: 44 (0) 20 8392-8585
FAX: 44 (0) 20 8392-9876
E-mail: Bushwd@aol.com.
Free postage in the UK. Europe: air mail at cost.
Try your bookstore first.

Contents

1
Introduction

The military aircraft as a platform for weapons. It certainly was not the way the system started out. But as the aircraft have evolved through the years, so too have the weapon systems that have been carried aloft.

The missions that came about from a platform on high have been varied and flexible, and have included both offensive and defensive systems. Then, too, there have been weapons to attack ground radar installations, along with decoy systems that can simulate an aircraft aloft to an enemy radar.

Of course, the first mission of the airplane in a war environment was that of observation. It would not be long, however, before the planes would be armed with primitive gun systems which allowed the adversaries to fire on each other.

Those systems would grow in effectiveness with machine guns mounted on the forward fuselage and firing through the propeller. As a defensive measure, many early bi-planes would also have a rear-firing installation.

Next came the first air-to-ground mission of the airborne platform with the capability added to drop bombs on the enemy. Crude at first, of course, but that capability would mature over the years, as well.

With the start of the Korean War, the airborne rocket would become a significant ground attack weapon, and sometime air-to-air weapon, being carried mostly by fighter and light-bomber aircraft. The rockets were usually fired in salvos against all types of ground targets with oft-times devastating results.

Air-launched missiles have asumed a high development priority in the late 20th and early 21st century. (Vadnais Photo)

Then came the first models of the airborne weapons of today—missiles and guided bombs. They represent a wide range of capabilities, and with all the major countries of the world using them, there is a constant technological battle to achieve the ultimate weapon.

This book will examine the history of airborne-launched weapons and a number of variations. It is an amazing story, where this

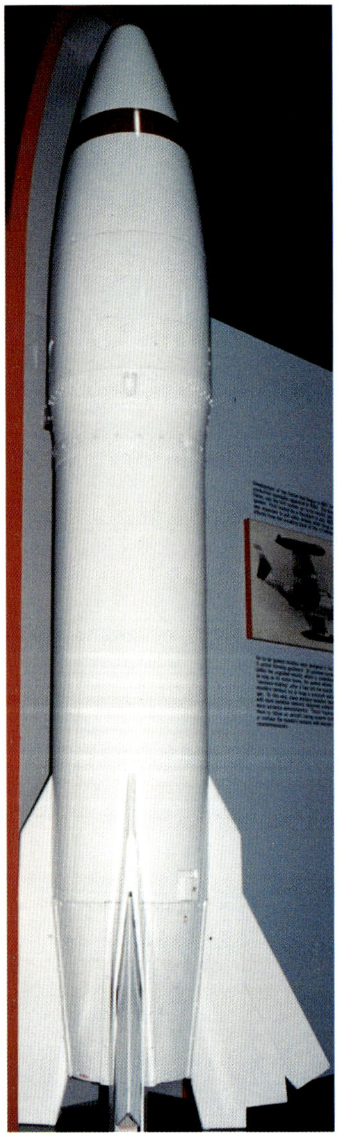

AIR-2A Genie nuclear missile. (Vadnais Photo)

A Navy A-4 Skyhawk fires an AIM-7 Sparrow at the China Lake test range. (U.S. Navy Photo)

technological base has moved in such a few short decades and where it may go in the future. Following is the layout of the book and what will be covered in each chapter.

Initial air-launched weapons were unguided rockets, but accuracy limitations of these systems would call for more advanced systems which could be guided directly to the target, either by the launching aircraft or by a guidance system onboard the weapon system.

Chapter 2 will cover the wide-ranging subject of U.S. Air-to-Air missiles (AAMs) which are enjoying a huge developmental priority worldwide moving into the 21st century. As the cost of fighter aircraft continues to spiral, the importance of protecting those fewer planes, and providing them with more effective offensive and defensive power capability, becomes higher and higher. The foreign developments in the AAM field are covered in Chapter 3.

With the increased capability of more effective AAMs, a single fighter can be as effective as a number of earlier systems. Destroying an ememy fighter out-of-sight of the pilot and many miles distant has now become a standard capability of the modern AAMs. No longer is it necessary for the launching aircraft to even be pointed toward the target.

With so-called off-boresight missiles, aircraft have the capability of firing "over the shoulder" at an attacking adversary. It is getting to the point that these sophisticated Air-to-Air missile systems are starting to have a mind of their own.

A pair of F/A-18s fire a salvo of AIM-9 Sidewinders. (U.S. Navy Photo)

During the initial flight test program for the F-22, test firings of internal stores were conducted. Here, an AMRAAM missile is on its way to the target. (USAF Photo)

Wind tunnel testing of an F-16 fighter carrying a pair of Sidewinder missiles on its wingtips. (AEDC Photo)

An Air-to-Ground Bullpup missile is shown under the wing of an F-100 fighter. (USAF Photo)

The AAM development design base, though, does not just exist in the United States, as a number of other countries, including the former Soviet Union, Israel, Great Britain, France, and Germany, among others, have also been big players in this strive for excellence. It should also be noted that AAMs are not mounted on just fighter-type aircraft, but also on rotary-wing helicopters.

Chapter 4 addresses the other main portion of the airborne weapon mission, that being air-to-ground missiles (AGMs), which also cover a wide range of missions, U.S. systems in particular. Unlike the air-to-air missiles, AGM missiles are mounted on both fighters and bombers, with some of the latter missiles employing airbreathing propulsion systems and being extremely large aircraft-size vehicles.

As was the case with the AAMs, there is also a world-wide interest in these systems, too, but not to the level of the air-to-air missiles. Foreign air-to-ground missile systems are covered in Chapter 5. Chapter 5 also covers the air-launched category of U.S. and foreign air-launched guided bombs. The differentiation between these weapons and the aforementioned Air-to-Ground missiles is that these systems do not possess a propulsion system as such. But they are guided by a number of different techniques, including control surfaces which can be altered to correct a final trajectory. This

A test Tancit Rainbow Air-to-Ground missile hangs below the wing of an A-6 Intruder preparing for a test launch. (U.S. Navy Photo)

A non-offensive air-launched vehicle is this Firebee drone, shown here under the wing of a C-130 transport. (USAF Photo)

Assuring that an aircraft can successfully launch its air-launched stores is accomplished by wind tunnel testing. The stores of an F-15 are shown here. (USAF Photo)

type of air-launched system enjoys continued refinement and development during the early 21st century.

Chapter 6 would have to be considered a "catch-all" chapter covering a number of other missions performed by airborne weapon systems. Decoy missiles are the main types of missiles addressed in this chapter, but there are a number of other interesting applications of this concept.

When you really get down to it, you might consider paratroopers as an off-shoot of the air-launch concept. Then, there is also the air launch of towed gliders, a concept that was used during World War II to great success.

2
U.S. Air-to-Air Missiles

This chapter provides the complete history of Air-to-Air missiles (AAMs). All the variants of each major missile type are covered, along with the additional capabilities acquired from the upgrades.

The chapter will also cover the Air-to-Air missiles developed from the earliest 1970s-vintage systems to the future systems, with unbelievable capabilities, that will be appearing in the next millennium.

The idea for Air-to-Air Missiles originated following World War II following the success of aircraft-mounted unguided rockets. At the time it was felt that development of such systems would probably be accomplished by the aircraft companies. But that did not always prove to be the case because of the electronic aspects of the new systems, and a number of pure-electronic companies such as Raytheon, Philco, and others would assume that role on some systems.

In the following decades, Air-to-Air Missile development in the rest of the world would challenge that of the U.S., with Great Britain, France, and the Soviet Union being the prime developers. A number of smaller countries, without the capability to develop their own systems, would be buyers of these systems.

One trend that was noted was the adoption of ground-launched AAM systems to an airborne launching pad. Such an example was the Stinger, which was initially designed as a land soldier-mounted weapon. Later, the system would be adapted as a weapon for helicopters.

The capabilities of the newer AAM systems tend to boggle the mind, partially taking the pilot out of the "dogfight" business. It used to be a situation where the pilot was required to line up his air-to-air missile with the target before the missile could be released.

Now, the missiles seem to possess minds of their own, and are able to accomplish off-boresight, over-the-shoulder launches at an unseen enemy. The world competition to develop the ultimate AAM is intense and currently enjoys a high priority.

The interest in development of Air-to-Air Missiles continues to spiral into the next millenium with the emphasis more on medium range capabilities than the shorter range missiles. Operation Desert Storm vividly illustrated the need and effectiveness of such systems. So here they are, the amazing family of U.S. Air-to-Air Missiles!

AIM-4 Falcon (USA)
Manufacturer: Hughes Missile Systems
Native Users: USAF, U.S. Navy,
Foreign Users: Sweden
Variants: AIM-4, 4A, 4B, 4C, 4D, 4E, 4F, 4G, XAIM26, AIM- 26A, B, AIN-47A, XAIM-4H. Launching Aircraft: F-89, F-102, F-106, and F-4
Operating Period: 1947-1971
Mission: To produce the first guided Air-to-Air missile

A size comparison of the Falcon missile. (USAF Photo)

The Story:
Compared with the technology of today's Air-to-Air missiles, the Falcon was a mere baby, but this first operational AAM set the techology base for the many systems that would follow.

The need for such a system was quickly addressed by the new U.S. Air Force in 1947. To that end, a contract competition for development of a guided AAM for the next generation of fighters was undertaken. A number of companies bid for the contract, but to the surprise of all, Hughes Aircraft won both parts of the overall contract.

But before the development of a guided system could be completed, it was necessary to develop the accompanying fire control system onboard the launching aircraft.

Hughes aggressively addressed the program, with a new production plant being built in Culver City, California.

Considering that this was the first AAM, the technology was surprising and excellently packaged. Using a Thiokol solid propellant rocket motor, the Falcon accelerated like a bullet with top loadings of up to 50 Gs. Steering was accomplished by control surfaces on the small wings. There were also small receiving antennae lo cated on the forward fuselage.

As far as the AIM-4's designation, there were a number of variants (A through D) which incorporated both IR and SARH guid

An artist's concept of the Falcon missile. (Hughes Drawing)

Loading a Falcon under the wing of an F-4. (USAF Photo)

Two early Falcon versions. (USAF Photo)

Falcon loading operations. (USAF Photo)

A Falcon is shown in front of the armament array for the F-106 air defense fighter. (Air Force Museum Photo)

The Falcon family of AAMs. (Air Force Museum Photo)

ance systems. The IR-type versions proved to be the most popular, since the missile could be launched toward the target and the launching aircraft could start its egress from the area. Almost 50,000 of those early versions were built.

The lengths of the early Falcons varied from 77.8 to 79.5 inches with launch weights from 110 to 134 pounds. The speed and range of all these variants were very similar at about Mach 3 and six miles, respectively.

The designation E was the next-in-line for the Falcon family, but this was an entirely new bird, so much so that it would be called the Super Falcon. Just about everything about the new system was new and improved, including a longer-burning motor, new radome, and a more powerful warhead.

Another new motor would come on line with the F version, along with an improved SARH guidance system. Most unique was an aerodynamic device in the form of a metal rod protruding from the nose, the purpose of which was to create a shock wave, thus increasing the flow and aerodynamic characteristics of the missile.

The AIM-4G took the F version and added another dimension with the addition of a new seeker, allowing the acquisition of targets at greater ranges along with the capability to lock on to smaller targets.

With the start of the 1960s, the Falcon world acquire a different designation, the XAIM-26. The experimental system was de-

AIM-4C Falcon. (Bill Holder Photo)

These AIM-4F Super Falcons carry a distinctive red and white paint scheme. (Bill Holder Photo)

Launching device for internally-carried AIM-4F Falcons. (Bill Holder Photo)

veloped in order to overcome the head-on attack problems that were noted in the then-current IR and SARH-guided systems.

The operational fall-out of the system was the AIM-26A, which would be the first true guided missile to employ a nuclear warhead, actually the same warhead as the Genie. The warhead was activated by four proximity fuzes. The weight of the AIM-26A had almost doubled from the original Falcon's 203 pounds, even though it was only seven inches longer.

The 26B version quickly followed, a version that would be licensed and built by Saab and called the RB-27. Two thousand of this variant were built, the most since the 2,700 of the AIM-4G. The 26B version would greatly increase in weight, up to 262 pounds.

During the tenure of the Falcon, the system would be tailored to new aircraft coming online, with one of the last examples being the proposed F-108 Rapier Fighter, a fighter which was to be used with the B-70 bomber.

The advanced Falcon would be called the GAR-9 (or AIM-47A) and would greatly increase in size to 126 inches in length, up from the maximum 86 inches of the previous models.

But the aspect of the new system that set it apart was its dramatic range of 115 miles, which was about 20 times greater than its Falcon brethren. The missile guidance plan called for it to use mid-course guidance, being directed by the ASG-18 radar, and then homing on the target with a terminal IR system. The advanced system, though, would never make it to the production phase.

The final Falcon carried the XAIM-4H designation, the only Falcon that would not receive a GAR designation. The system carried an advanced proximity fuze, but the program would be terminated in 1971.

The newer systems were then starting to pass the Falcon as it faded into obscurity. But it will always have the distinction of being the first Air-To-Air missile.

This Falcon has a distinctive black nose and red secondary tailfins. (Scott Vadnais Photo)

AIM-26B nuclear Falcon. (Scott Vadnais Photo)

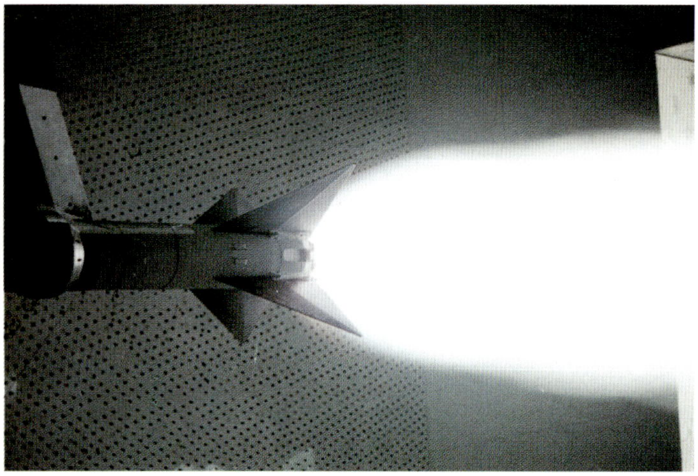

A propulsion system test of an early Sparrow in the AEDC 16-foot transonic wind tunnel. (USAF Photo)

AIM-7 Sparrow (USA)

Manufacturer: Sperry Gyroscope, Douglas Aircraft, and Raytheon
Native Users: USAF, U.S. Navy, USMC
Foreign Users: Egypt, Israel, Japan, Saudi Arabia, Canada, Greece, Taiwan, Turkey, Malaysia, Thailand
Variants: AIM-7A (XAAM-N-2/AAM-N-2), AIM-7B (AAM-N-3), AIM-7C (AAM-N-6), AIM-7D (AAM-N-6A, AIM-101), AIM-7E, AIM-7F, and AIM-7M
Launching Aircraft: F5D1, F-4, F-14, F-15, F-16, F-18, Arrow (Canada), F-104S (Italy)
Operating Period: Early 1950s-present
Mission: To develop a large Air-to-Air missile system for the next generation of fighter aircraft.
The Story:

The second Air-To-Air missile to carry a "bird" nametag, the Sparrow Missile would be a longstanding system, mounted on a number of U.S. and foreign fighter aircraft. The unique aspect of this system was that there were three distinct families, along with a different contractor for each family—Sperry, Douglas, and Raytheon.

An AIM-7M Sparrow races away from an F-15. (USAF Photo)

AIM-7M launch from an F/A-18 Hornet. (Raytheon Photo)

Sperry started with the program immediately following World War II under a U.S. Navy contract. The program was known as Project Hot Shot. Two years later, the company would acquire the initial Sparrow contract, ie the XAAM-N-2 Sparrow I. It would only be another two years before the vehicle made its first flight test.

The missile was guided by a simple beam rider system, where fuselage-mounted antennae received commands from the launching aircraft. The commands were then relayed to the four cruciform wings, which were pivotable and could interpret the commands into missile movement. The tail fins were fixed and provided stability for the missile system.

Propulsion for the this Sparrow was provided by an Aerojet solid motor. Declared operational in 1958, the Sparrow 1 was soon operating with both U.S. Navy and USMC units with both the Pacific and Atlantic fleets.

The smallest of all the Sparrows, the Sparrow I was 140 inches in length, had a 39-inch wing span, and a 110 pound launch weight. Range was approximately five miles. Some two thousand of the model were produced, with the first launch occuring in 1988 from an F-16A.

The Sparrow II (AIM-7 or AAM-N-3) was the product of the second contractor, Douglas, a version that was specifically planned as the main weapon for the F5D-1 Skylancer. Due to a disagreement between the Navy and Douglas concerning the type of guidance system to be used—either SARH or a fully active system—the Navy decided to cancel the program in 1956.

But the Sparrow II was exactly what the Royal Canadian Air Force had been looking for, and they adopted it for its Arrow supersonic fighter. The system did not work in this application, either, and the experiment was terminated in 1958. Interestingly, the Arrow program met exactly the same fate.

The Sparrow II was slightly larger than the first bird, being four inches longer, but weighing considerably more at 110 pounds.

In 1955 the third contractor, Raytheon, had started design and development work on the (AAM-N-6) Sparrow III system. The external dimensions were similar to the Sparrow II, but it had shed 40 pounds and sported an impressive range of 25 miles. Guidance was now a Continuous Wave SARH system.

Mounting details of a pair of Sparrows on the fuselage of an F-15 Eagle. (McDonnell-Douglas Photo)

An AIM-7 stands alone. (Scott Vadnais Photo)

The lightening of the Sparrow III came from the use of a light metal alloy for the body. The III also carried an Aerojet solid propulsion system and would become the AIM7C, reaching operational status in 1958.

The AIM-7D (originally the AAM-N-6A) would be nearly identical to the C version, with the exception of the substitution of a Thiokol liquid pre-packaged rocket motor. It would be in this time period that the highly effective system would be adopted by the Air Force (called the AIM-101) to carry on its F-4C Phantom fighters, interestingly another Navy development. The F-4s carried four Sparrows contained in an internal weapons bay.

Depending on the particular application, a number of different illuminating radars were used with the AIM-7D. The AIM-7D would retain the same external dimensions as the 7C, but would add on considerable weight, up to 440 pounds.

The third propulsion system, with the Rocketdyne solid propellant motor gave the E version (originally the AAM-N-6B) punch with a burnout velocity of almost Mach four. A lethal continuous rod warhead also added a huge increase in the power of the missile system. The E versions would see action during the Vietnam War and were the most-produced, with 25,000 being built.

The Hercules Mk 58 solid propellant motor was a significant change for the AIM-7F version, providing the capability to carry an 88-pound warhead. In addition to Raytheon, General Dynamics also built a number of this version, which gained 12 pounds and was rated as effective out to 28 miles.

Both contractors would then revert to building the M version, which had an inverse-processed digital mono-pulse seeker head providing a range out to an impressive 62 miles, the same figure quoted for the F version.

The AIM-7M has a number of unique capabilities, including the capability to intercept both low and high altitude targets, along with aircraft and limited missile capabilities flying at supersonic speeds, intercept of targets in severe countermeasure environments, engagement of highly-maneuverable targets in both forward and rear hemispheres, intercept in snap-up and shoot-down conditions, and intercept of closely-spaced multiple targets.

The AIM-7M program aimed at reducing cost and improving performance in the ECM and look-down clutter regions. Production started in 1980 and entered operational service in FY 1983. It would be carried by most of the current family of Navy and Air Force fighters.

The next Sparrow, the AIM-7P/RIM-7P built by Raytheon, could well have reached the total potential of the system. Using a semi-active/active radar guidance system, the P version of the Sparrow carried the same external dimensions as the AIM-7M, but came in slightly heavier at 514 pounds.

The AIM-7R started development in 1990. It differs from earlier models in that a majority of its maneuverability is acquired from its aft tail fins. The change occurred so that the system could carry more fuel.

Over 65,000 AIM-7s of all variants had been produced through the mid-1990s.

This early artist's concept of an F-16 shows a pair of Sidewinders mounted on the wingtips. (General Dynamics Photo)

A measure of the longevity of the Sidewinder family is illustrated by this F-100 launch of an early version. (USAF Photo)

AIM-9 Sidewinder (USA)
Manufacturers: Loral Aeronutronic, Raytheon, Hughes, European Consortium
Native Users: USAF, U.S. Navy, USMC
Foreign Users: Nationalist China (first wartime use), Royal Air Force, German Air Force
Variants: OAR-8, AIM-SIB, 9C, 9D, 9E, 9G, 9H, 9J, 9L, 9N, 9M, 9X
Launching Aircraft: F-4, A-7, F-8, A-98, F-111, F-14, F-15, F-16, F-18, F-22, Jaguar, Buccaneer, and others
Operating Period: Late 1950s to present
Mission: To master the required technology and develop the first infra-red guided Air-to-Air missile
The Story:
Before beginning this amazing story, it is really necessary to recognize the accomplishment of the AIM-9 Sidewinder and all it has done through its four decades. Quite frankly, it is without doubt the most famous air-launched missile in the world. And the story is far from over, as more advanced versions continue to be proposed and produced.

It all started in 1949 with a low-budget effort and a small research team at the Navy's China Lake facility. The effort was directed at determining the feasibility of an infrared guidance system. Much of the effort was devoted to the development of a small seeker head, not an easy job in the period before mini-electronics.

The first actual rounds, though, would be fired only four years later. The names were a confusing montage, with the first firing using a XAAM-N-7 designation. The Air Force would call the first

Late version Sidewinders on an F/A-18. (U.S. Navy Photo)

This USMC F/A-18 includes Sidewinders in its weapons suite. (U.S. Navy Photo)

production version the OAR-8, and it would achieve operational capability in 1956.

The early Sidewinders were far from being high tech, with aluminum tubing being used for the fuselage. The shape of these early birds, though, would basically establish the size and shape of Sidewinders to follow.

First, there were the two sets of fins; the moveable forward set provided the aerodynamic forces to turn the missile, while the fixed rear fins provided the stability for the missile. The first versions were capable of accelerating to Mach 2.5.

An interesting innovation on the rear set of fins was a so-called rolleron, actually a small gyro, one of which was located on the tip of each rear fin.

Of course, the key to the success of the Sidewinder in all its configurations to follow was its front seeker, which seemed to have a mind of its own. Its purpose was to lock on to the IR signature of an enemy aircraft and then provide the guidance directions to direct the missile to the target.

It all sounded easy, but there were certain conditions that made the task difficult on occasions. The effectiveness of the system was greatly compromised by poor visibility, cloud cover, or rain, making acquisition more difficult.

There was also a tendency, on occasion, to lock onto the largest IR source on earth, ie the sun. Then, there were also ground objects that brought about confusion, like the sun reflecting off the surface of a lake or pond.

The method of releasing the Sidewinder to do its job is well documented and has remained basically unchanged through the years as the system has evolved.

First, the pilot would energize the homing head after pointing his launching aircraft toward the potential target. There would then be a growl from the system once lock-on had been achieved. As the target got closer, the sound would intensify until the missile was released to do its deadly job. Then, there would be no further asso-

This photo sequence shows the launch of an AIM-9M Sidewinder from the YF-22 prototype. (Lockheed Photo)

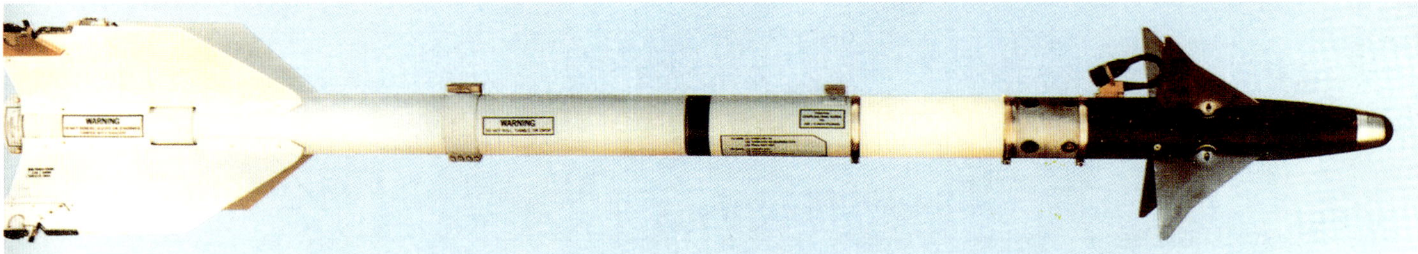

The basic shape of the Sidewinder system has remained basically unchanged through its long evolution. This drawing shows a later-version S model. (Raytheon Photo)

ciation with the pilot or launching aircraft, which allowed the plane to move toward safer pastures.

In 1962, the terminology for U.S. Air-To-Air missiles would change with the AIM designation coming online. Hence, the early Sidewinders would acquire the famous AIM-9 designation, along with the A, B, and C desigations being established. These first three variants actually were little different from the original test version, with a few exceptions.

AIM-9 on an F-16. (Scott Vadnais Photo)

The B version was the most popular and was produced in huge numbers, almost 81,000 being produced by Philco Ford, Raytheon, and a multi-company European consortium that introduced one change in the system, substituting a silicon nosedome in place of the normal glass unit. It carried a burn time of about 20 seconds and a minimal range of only two miles. A total of 1,600 of the B version were built.

Carrying a new Rocketdyne Mk-36 propulsion system, the burn time of the C/D variants was tripled to a full minute, with a resulting increase in effective range to 11 miles. The maneuvering capabilities of the systems also benefitted from the lengthening of the fin span by two inches, up to 24 inches.

Production duties for the AIM-9C and 9D were split between two manufacturers, Motorola building the 9C, and the 9D being built by Ford. There was a significant difference in the guidance

The Sidewinder continues to be a part of the AAM inventory. Here, in this artist's concept, the missile is shown being launched from the new F/A-18E/F fighter. (McDonnell Douglas Photo)

Testing of a Sidewinder AAM in an AEDC wind tunnel. (AEDC Photo)

AIM-9 on a rack and prepared for loading. (Scott Vadnais Photo)

A USAF Crew Chief makes final adjustments on a Sidewinder Missile before a mission. (USAF Photo)

systems, with the introduction of a Semi-Active Radar Homing (SARH) system which was designed specifically to be compatible with the APQ-94 radar of the Navy F-8 Crusader. The system, though, did not provide the performance desired and was discontinued.

The AIM-9D proved to be a very successful variant, and much of the new technology of the version would find its way into following versions. A new seeker dome provided considerable advantage over the previous models. The firepower was greatly increased with a potent new 22-pound warhead, more than twice the weight of the previous ten-pound models.

A new Ford-designed seeker head added greatly to the capability of the AIM-9E, providing greater tracking capabilities. The E's production of five thousand units would be the most since the massive 9B production total.

The AIM-9G came next, and the burn time was again at the 60 second figure, along with an intercept range of 11 miles. Improvements were also made to the seeker head, with a significant production run of 2,120.

The AIM-9H, with a production run of 7,720, brought improved tracking speeds and solid-state electronics to the model. There were also the new double-delta control fins which provided greater turning rates than previous Winders. The H also brought some minimal off-boresight capabilities to the missile.

If production numbers were any indication of the effectiveness of the different Sidewinder versions, then the J and L versions must have been killers, with 10,000 and 11,700 being constructed, respectively.

The J was actually a rebuilt version of the earlier B version. Burn time was reduced 20 seconds from the H version, derived from a different motor burn pattern with a resulting greater overall acceleration. There were both J-1 and J-3 variants of the version.

The 9L started production in 1977 and could well be considered as the first of the third generation of the model. Externally, the

AIM-9 on an F-104 Starfighter. (Scott Vadnais Photo)

AIM-9 on an F-10 Starfighter. (Scott Vadnais Photo)

An F-15 pilot checks out a wing-mounted AIM-9. (USAF Photo)

system was characterized by new pointed guidance fins. Other new innovations included a new high-tech fuze which featured laser diodes.

As the letters continued, the M version would evolve as an improvement of the L variant with increased ECCM with increased ECCM capabilities. The Raytheon-built version was used extensively on USAF fighters.

The AIM-9N was actually the new designation for the earlier J-1 version with an impressive production run of 7,000.

The AIM-9P, built by Ford Aerospace, was modified from conversions of the B/E/F versions, along with new production. With 13,000 being produced, the system introduced increased reliability and a new motor, which produced considerably less smoke, as well as a new warhead fuze. There were at least two variants to the P version, the P-3 and P-4. The P-3 demonstrated increased lethality. The P-4, which was developed for foreign military sales, had a new guidance system and an increased acquisition range.

The final, and possibly the ultimate Sidewinder, was evolving at press time under the name of the AIM-9X. The goals of this new system are ambitious to say the least, since it is hoped to have the capability to shoot down low-observable cruise missiles that currently have the ability to elude the best Air-To-Air missiles.

When compared to the latest M and P versions, the 9X will have significantly greater capabilities, with better maximum and minimum ranges, higher maneuverability, and better clutter and countermeasures capabilities. Preliminary plans called for some ten thousand of the model.

The need for this advanced version was brought about because of the projected capabilities of the Soviet AA-11.

Since the new missile will be the prime Air-To-Air missile used with the F-22, it may be necessary to reduce the fin size in order to fit into the weapons bay of the new fighter.

It takes a trio of USAF crewmen to carry an AIM-9 to its waiting fighter. (USAF Photo)

The installation of an AIM-9 an an AV-8 Harrier is clearly visible from this view. (USAF Photo)

In December 1996, the AIM-9X contract was awarded to Hughes, surpassing competing bids from Raytheon and British Aerospace. But since Hughes was later acquired by Raytheon, it could be stated that Raytheon ended up as the overall contractor. Initial plans called for the production to commence in 1998 with IOC in 2002 and production completion in 2011. A total production has been estimated as high as ten thousand missiles.

Dimensionally, the missile was quoted at 9.5 feet in length, only five inches in diameter, and carried a 21-pound payload. Range is to be two miles with a 50,000 foot altitude capability and a higher velocity capability. Propulsion is provided by Thiokol with its MK 36 solid motor.

Unfortunately for the USA, the AIM-9X came up short in a competition to arm Australia's F/A-18s, losing to the French ASRAAM system. However, the system will still find considerable use with U.S. services. As of mid-1999 the program was in flight test by the Navy at China Lake, California, with significant buys still planned.

AIM-54 Phoenix (USA)

Manufacturer: Hughes, Raytheon
Native Users: U.S. Navy
Foreign Users: Launching Aircraft: F-111B, F-14A-D
Operating Period: 1960 through present
Variants: AAM-N-11, AIM-54A, AIM-54B, AIM-C, AIM-C+
Mission: To produce a wide-area defense Air-To-Air Missile far surpassing any missile to date
The Story:
For the time period when the Phoenix was developed, this system was the most effective Air-To-Air Missile system in the world. Its huge range was reported as covering an area over 12,000 square miles, and from altitudes from sea level up to the altitude limits of any enemy aircraft or missile.

The Navy missile was initially designed to arm the canceled Navy F-111B, but would be adopted for the F-14 Tomcat and its variants, and that is still the case for the system in the late 1990s.

Amazingly, this program has been around for parts of four decades, with the program actually beginning as a Navy program with Hughes Aircraft in 1960. The program initially carried the AIM-N-

Phoenix missile being launched from an F-14. (U.S. Navy Photo)

AIM-54A Phoenix with a cross section internal view. (Scott Vadnais Photo)

11 designation and was directed toward the replacement of the AIM-47B missile system.

The initial version of the system was impressive with a range of 150 miles. In order to aid in close-in target identification, the system was assisted by an IR tracker.

The capability of this sophisticated system was far beyond the state-of-the-art with the ability to fire six missiles and engage six different targets simultaneously.

The Mk-47 Hercules solid motor, or Aerojet Mk-60, was the potent propulsion system used with the AIM-54A with a semiactive/active radar guidance system. It would be the type of system used in all of the Phoenix variants. The system was also fitted with a large proximity-fuzed high-explosive warhead.

The Phoenix launch aircraft must be carrying the AWG-9 weapon control system, which has the capabilities to lock on to not only high-performance aircraft, but also cruise and air-to-surface missiles. The first AWG-9 was fitted to an F-14A fighter in Feburary 1970.

The Phoenix flight test program for the system's initial version began in June 1973 and showed an 88 percent success rate of the 56 missiles that were launched.

All of the Phoenix variants would sport the same large dimensions of 13 feet in length, a three-foot maximum span on the fins, which were triangular in shape and stretched to about half the length of the body, and a body diameter of 1.3 feet. The initial version would weigh 985 pounds, with the B, C, and C+ variants close to that weight.

The long-burn capability of the propulsion system provided the missile with a burnout velocity of almost 4Gs. Then, combined with the extremely large hydraulically-driven control surfaces on the fins, the Phoenix was endowed with high maneuverability capability.

Flight testing for the Phoenix began in 1965 and immediately demonstrated its significant capabilities. The system far surpassed existing missiles at the time, setting a number of new records for distance to kill a target. It demonstrated the capability of accomplishing a direct hit on a target over 83 miles from the launching point.

Production total for the A version was over 2,500 units with the period of production lasting from 1973 until 1980.

The AIM-54A first entered operational service in September 1974 when two squadrons of Phoenix-equipped F-14s were deployed on the USS *Enterprise*.

That initial version, though, would be replaced with the similar AIM-54B version, which incorporated several improvements, including substitution of metal fins in place of the homeycomb construction of the initial version.

Along with improvements to the hydraulic and thermal conditioning systems, the B version also topped the thousand pound mark at 1,020 pounds. The B version entered service in the late 1970s.

Then came the C version, which could be considered a practically-new missile system. It was laced with modern technology with the introduction of new digital electronics, a solid-state radar, and an all-inertial reference unit.

Flight testing of the AIM-54C would commence in 1980 with full production starting in 1982. With a 13-foot body length and 1.25 foot body diameter, the AIM-54C weighed 985 pounds.

A modification would follow the C version, but it would not receive a new letter designation, instead being denoted as the AIM-54C+ version which would incorporate advanced solid-state electronics.

AIM-120 AMRAAM (USA)
Manufacturer: Hughes Missile Systems, Raytheon Company
Native Users: USAF, U.S. Navy
Foreign Users: Germany, United Kingdom, France, Bahrain, Turkey, Korea, Norway, Netherlands, Sweden
Variants: AIM-120A, B, C, and D
Launching Aircraft: F-4, F-14, F-15, F-16, F/A-18, F-22, Harrier, Toronado
Operating Period: 1982-present
Mission: Designed as a replacement for the AIM-7 Sparrow, AMRAAM has a mission as an all-weather, all-aspect, look-down, shoot-down missile designed to counter sophisticated air combat threats.

AMRAAM launch from an F-16. (Raytheon Photo)

The Story:
It probably would come as a surprise to learn that the seemingly-new AMRAAM (Advanced Medium Range Air-to-Air Missile) began its design lifetime in the early 1980s when Hughes Aircraft signed the prime contract, with Raytheon later being denoted as the follow-on contractor. The original competition had five companies vying for the contract.

The system actually changed missions early in its design phase, being initally designed as a Beyond Visual Range program with the limited goal of replacing the Sparrow. But that mission would be broadened as the medium range AAN NATO requirement, and would acquire its AMRAAM name.

In a wide-reaching Memorandum of Understanding, signed by the United Kingdom, West Germany, France, and the USA, it was agreed that the United States would develop the AMRAAM, while the United Kingdom and West Cermany would develop the Sidewinder replacement, to be coined the ASRAAM.

The initial IOC of the system had been planned for 1986, but technical problems slipped the date back three years. The high-tech system would finally achieve that goal in 1991. Raytheon completed production qualification in January 1984, after successfully completing live-fire testing of five qualification lot missiles, the final two of which were direct hits.

As the missile system neared IOC, the AMRAAMs that were being manufactured by both companies were achieving an impres-

Length	3.65 m (144 in.)
Diameter	178 mm (7 in.)
Wing Span	533 mm (21 in.)
Control Surface Span	635 mm (25 in.)
Weight	156.8 Kg (345 lbs)
CG	2.04 mm (80.4 in.)

AMRAAM characteristics. (Raytheon Photo)

One of the YF-22 prototypes launches an AMRAAM during the plane's initial flight test program. (Raytheon Photo)

AMRAAM launch from an F/A-18. (McDonnell Douglas Photo)

AMRAAM launch sequence from an F-22. (USAF Photo)

sive success rate of over 80 percent. As an example, in May 1990, four AMRAAMs were launched from a single F-15 Eagle fighter against four targets in an extremely heavy electronic countermeasures environment. Three missiles hit their targets, and the fourth passed close enought that it would have destroyed its target had it been armed with a warhead.

As of 1998, an inventory of more than 12,400 of the potent AAMs were planned for USAF and the USN alone. To date, there have been three versions of the AMRAAM, with the A version first delivered by Hughes in 1988 to the USAF 33rd Tactical Fighter Wing. The improved B and C versions were both in production during the late 1990s. The B version is operational with RAF Harriers and Tornados.

The AMRAAM would also be part of a Raytheon proposal in the Eurofighter competition with a European consortium. The system was called the AIM-120B Plus, an improved version of the AIM-120. Another missile using AMRAAM technology was the so-called Extended Range Air-to-Air Missile, which proposed the use of a dual-pulse solid rocket motor. Operational status for such a system was estimated to be about 2005.

In a late development, the U.S. government allowed the AMRAAM to be a part of the F-16 package to Bahrain. Around the turn of the century, the Navy was considering an AIM-120D version which would use technology from the canceled AAAM program.

Without doubt, the technology of the AMRAAM was a dramatic step forward over previous air-to-air systems. It provided the capability to attack low-altitude targets. A number of AMRAAMS can be fired simultaneously at multiple targets, along with having the capability of performing evasive maneuvers.

Versatility was a goal in the AMRAAM design, with the capability to mate with a number of U.S. and foreign aircraft systems. Undoubtedly, as the system continues to evolve in the next millennium, it will become compatible with numerous other aircraft. There

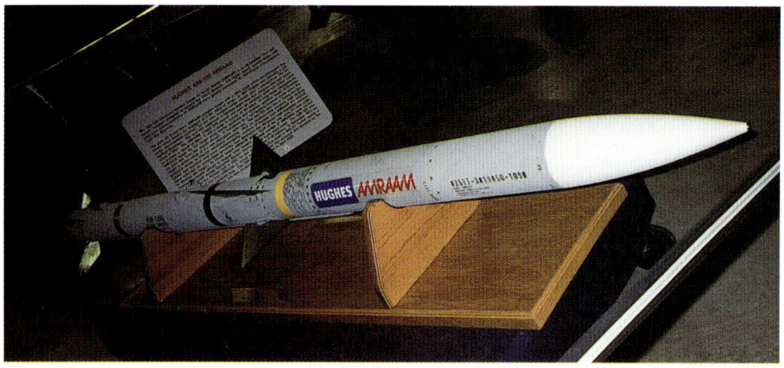

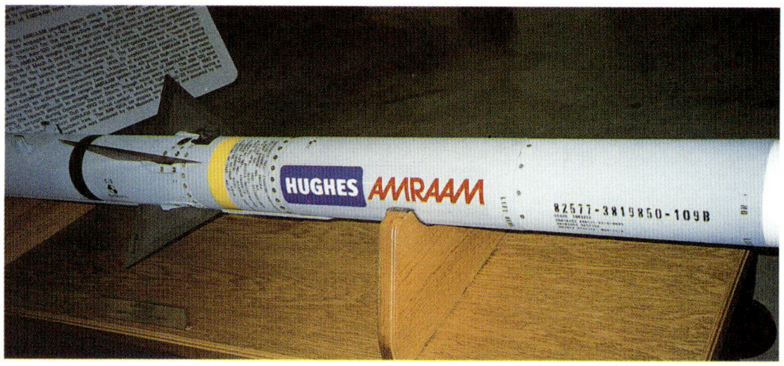

AMRAAM display at Air Force Museum. (Bill Holder Photo)

is little doubt that it will become the Sidewinder of the next decade, if not longer.

Looking at the parts and pieces of this system, the AMRAAM has a slender cylindrical body with a sharply-pointed nose. The missile features a pair of fins, one set at mid-hody and a tail set alligned with the forward set. The tail control provided by those rear fins and a highly responsive autopilot add much improved maneuverability to the AMRAAM. The body is an amazing 12 feet long, with a diameter of only seven inches.

The guidance system is an inertial and command intertial system with active radar terminal homing. Located on the front of the 335-pound missile is a high-explosive directed-fragmentation warhead weighing about 50 pounds. Initially using a Hercules solid motor in the early versions, the current AMRAAMs use an Aerojet two-stage solid-propellant rocket engine. The engine provides quick reaction for minimum-range intercepts and high energy for extended range kinematic performance.

The AMRAAM contains a high-power, active radar seeker operating in the X-hand with a programmable pulse-doppler waveform to enable opeation in all weather while providing acute visibility in heavy clutter environments. The monopulse seeker features highly-selected discriminants in range, doppler, and angle, rapid automatic gain control, and digital signal processing to improve performance in all environments.

This special trailor is used to carry AMRAAMs about a hangar prior to loading, on F-15s in this case. (Scott Vadnais Photo)

The AMRAAM launching aircraft has a track-while-scan capability in order to facilitate the multiple target capability. Launch data is provided to the pilot by the aircraft's computer. When the pilot makes the decision to launch, the position data of the target and that of the launching aircraft is input.

Computer-generated drawing showing a full load of six AIM-120Cs in F-22 Main Weapons Bay. (USAF Photo)

A close-up view of a wing pylon mounting on an AMRAAM missile. (Scott Vadnais Photo)

The business end of an AIM-120. (Scott Vadnais Photo)

AIM-120 ready for loading on a 2FS, 325FW F-15, Tyndall AFB, Florida. (Scott Vadnais Photo)

Once launched, the AMRAAM quickly accelerates to Mach 4, with the inertial autopilot providing the data for the intercept. Once in the close vicinity of the target, the AMRAAM's terminal homing guidance comes into effect and locks onto the target.

The ultimate test, of course, of any air-to-air missile system is actual combat. The AMRAAM can put three notches on its gun belt, as it accomplished two kills in Desert Storm and a single victory in Bosnia.

Confidence in the system was verified in the late 1990s when the USAF levied three different contracts for additional AMRAAM production. The first contract was for 483 units through 2001. 180 were to be delivered to the Air Force and 100 to the Navy. The remainder were to be delivered to Israel, Italy, Japan, Spain, and Turkey through the Foreign Military Sales Program.

There are strong possibilities of future upgrades for the AMRAAM. Areas considered are advanced counter-measure capabilities, improved software, and an improved propulsion system.

A recent application of the AMRAAM is in a ground-launched application where five were mounted on a HUMVEE vehicle.

Another off-shoot of the AMRAAM system is the so-called BVRAAM (Beyond Visual Range Air-to-Air Missile) requirement. The system would use a dual-pulse rocket engine and was offered as a possibility to fulfill the British requirement for an AMRAAM follow-on in 2007. A ramjet version of the AMRAAM was also considered.

Still another variant of the AMRAAM is the Future Medium-Range Air-to-Air Missile (FMRAAM), which would employ a ramjet propulsion system developed by Aerospatiale Missiles.

This F-16 carries a full load of ordnance under its wings, including AIM-120s on the wing tips with AIM-9s and AGMs closer to the fuselage. (USAF Photo)

This F-16 is carrying a pair of AIM-120 missiles on the end of each of its wings. (USAF Photo)

A crewman secures AIM-120s before use. (USAF Photo)

Future U.S. AAM Missiles

Air Superiority Missile Technology System (AMST)

In the late 1990s, the Air Force's Wright Laboratory undertook the so-called AMST program, which was a study of high payoff, air-to-air missile technology for both existing and the next generation of AAM systems.

The program started in 1997 and is expected to continue through 2002. The goal of the program is to develop a highly-maneuverable missile to perform both medium and short-range missions, a system that could make many current systems obsolete. Of course, such a system would also have attractive export possibilities.

Artist's concept of the canceled Eagle AAM. (Grumman Drawing)

It was also revealed that the expected enhanced maneuverability of the proposed system would enable the system to be employed in both stand-off and dogfight applications.

The program is built on experience acquired from the earlier Alternate Control Technology (ACT) program, which demonstrated a hybrid reaction jet/tailfin control package that permitted flight angles-of-attack of up to 70 degrees.

Significant Canceled U.S. Air-to-Air Missiles

Eagle AAM System

In 1957, the Grumman and Bendix companies combined for a winning proposal for the so-called Eagle air-to-air missile, which was designed as a long-range air defense missile for Navy fleet defense at all altitudes. The missile was identified with the AAM-N-10 designation and was to be employed on aircraft some distance from the ships it was protecting.

Bendix had the prime systems responsibility for the entire system, while the launching and support equipment was the responsibility of Grumman. The solid propellant propulsion system was to be developed by the Aerojet-General Company.

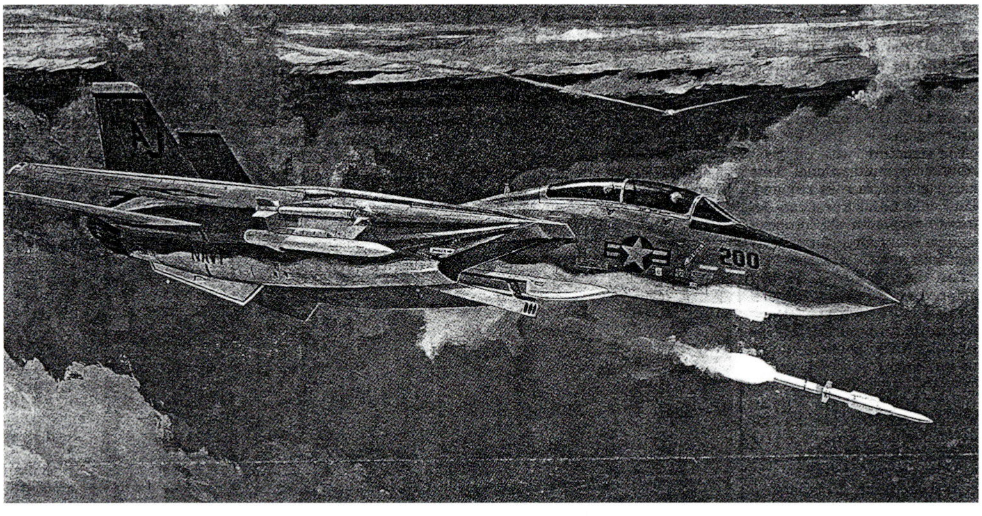

The canceled AAAM system is shown in this artist's concept being launched from a Navy F-14. (General Dynamics Drawing)

A unique F-15 carries the anti-satellite system to altitude for a launch. (USAF Photo)

Agile AAM Missile System
The Agile Missile was a 1970s joint USAF-U.S. Navy air-to-air missile program that reached the prototype stage before cancellation. Hughes was in charge of the guidance work, with the test program carried out at China Lake, California. The short range missile was designed to possess a significant off-boresight capability with control achieved by either thrust vector control from its solid-propellant motor, or from its control surfaces.

An annular fragmentation warhead was planned, to be detonated by an optical fuze. The missile was planned for use on the A-6, A-7, F-4, F-14, and F-15.

AIM-82A AAM System
The aforementioned Agile missile was actually brought about partly because of the cancellation of an Air Force-only dogfight missile with the designation AIM-82A. Little released information is available about this system.

Bomber Defense Missile
The Air Force also studied a Bomber Defense Missile which was to be carried by the B-1 bomber. It is estimated that this undefined missile never made it beyond the study stage.

Brazol Pve Arm System
Another joint USAF/USN AAM program that never saw the light of day was the Brazol Pve Arm system, which was intentionally designed to address the threat generated by the Soviet MiG-25 Foxbat fighter. The system was to have incorporated a radar homing guidance system. It is thought that the system would have used an airframe very similar to the Sparrow, with Hughes responsible for the guidance system.

Claw Missile System
Air-to-air combat experience has many times led to the development of new defensive systems. Such was the case of a 1974 dogfight missile looked at by the Air Force. The goal for the so-called Claw missile was to make it inexpensive and lightweight. Another idea that did not make it.

AIM-97A Seek Bat
The General Dynamics AIM-97A Seek Bat was also directed toward the powerful MiG-25. It reached the prototype test stage with a number of launches being attempted. The lineage of the missile was interesting in that it used the airframe of the air-to-ground AGM-78 missile system. It was also to use a solid-propellant motor and infra-red guidance.

AIAAM System
The AIAAM (Advanced Intercept Air-to-Air Missile) was designed in the late 1970s as an in-house Raytheon Project designed to eventually replace the AIM-54 Phoenix missile. The fuselage mounted both forward wings and rear control surfaces. The propulsion system moved away from the accepted solid fuel motor with either a hybrid or ramjet propulsion system considered, with oxygen being introduced to the powerplant by an underbelly supersonic inlet.
A Navy evaluation program was begun by the Navy in 1982, which consisted of simulations that might have eventually led to hardware, an eventuality which never took place.

An actual Anti-Satellite vehicle is on display at the Air Force Museum. (Bill Holder Photo)

AAAM System

The Advanced Air-to-Air Missile (AAAM) was a U.S. Navy program that was addressed by a pair of aerospace teams, one combined a Hughes/Raytheon team which competed against a Westinghouse/General Dynamics team.

Like the AIAAM system, the AAAM was also designed to be a Phoenix replacement. The goal was to make the system considerably lighter than the AAAM, allowing an F-14 Tomcat the capability to carry eight of the missiles.

Each AAAM team addressed the mission in an entirely different method, with the Hughes/Raytheon team proposing a dual-mode active-radar/IR guidance system and an advanced ramjet propulsion system.

The Westinghouse/GD group proposed a semi-active homing system with pure rocket propulsion. Plans called for demonstration and validation to occur in 1992 with full-scale development occuring in the mid-1990s. There were optimistic projections that as many as 4,000 of the missile type could have been built if the type had won the competition and gone into production.

Even though the AAAM would have been compatible with the F-15 and the future F-22, the Air Force showed no interest in the system. The Navy also considered the missile with possible applications to the F-14, A-12, and the A-6 Intruder. Cancellation of the program occurred in the early 1990s.

ARM AAM System

During the same time period, the Air Force also proposed the design of the ARM (Anti-Radiation Missile), which was planned to be used against enemy AEW aircraft. Although the program never reached fruition, there was the suggestion that collaboration in its development with the NATO countries might have been well served, but it did not happen.

ASAT System

During the early 1980s, one of the most interesting AAM design efforts was undertaken with the Air Force awarding a contract to the Vought Corporation for the development of an air-launched ASAT (Anti-Satellite) system.

The system consisted of an advanced interceptor missile which did not carry a warhead. The boost-to-orbit was to be accomplished by two stages of solid propulsion after it was released from the wing pylon of an F-15 launching aircraft. The system was to kill the enemy space object by actual impact.

Short Range Air-to-Air Missile System (SRAAM)

The requirement for such a short range missile came from combat experience in the Vietnam War, which highlighted the need for a high-acceleration missile for dog fight scenarios.

The goal of the SRAAM was to possess a short minimum range with the target maneuvering to its design limit. The missile was to be visually aimed and guided by a passive infra-red homing system.

3
Foreign Air-to-Air Missiles

Although Air-to-Air missile development probably started in the United States, in the decades that followed, many countries around the world have developed significant capabilities.

Great Britain, France, and the Soviet Union have been the prime foreign AAM developers, but new systems have also been developed in Israel, Japan, China, Brazil, and others.

Brazilian Air-to-Air Missiles

Piranha

The Piranha Air-to-Air missile was native-designed for the Brazilian Air Force and its Mirage III, F-5, and later for the AMX fighter.

The program was initiated in the mid-1970s, but was stopped on several occasions, stretching out the program before production was begun in the late 1980s. In an interesting development, the Brazilian Army and Navy, which had initially started the development, withdrew in 1977 with the program taken over by the Air Force until 1982.

The priority of the system was indicated from the fact that a new company, Orbita Sistemas Aerospaciais, was formed to ferry the system into the production phase.

The Piranha is a short range, fire and forget, dogfight missile. Its airframe is very similar in appearance to the U.S. Sidewinder. It is powered by a single-stage solid-propellant rocket motor with an infrared seeker, with the control achieved by the fuselage-mounted canards. A proximity fuze ignites the 26-pound fragmentation warhead.

British Air-to-Air Missiles

Through the years, the British have been one of the world's top Air-to-Air missile developers. A number of the country's developments have met with popularity with other countries and have been used to equip the air forces of many of those countries. The British have also been involved in a number of joint-country international programs.

Following are the significant developments from the British aerospace industry.

Hawker Siddeley Firestreak

The Firestreak Air-to-Air Missile was in service with the RAF and Royal Navy during the 1960s. The infrared-guided system armed the Sea Vixon and Lightning aircraft.

Powered by a solid-propellant rocket engine, the Firestreak had a metal cylindrical body, along with cruciform wings and tail surfaces. The guidance unit was located behind the radome, with control maintained by a proportional navigation system. A highly-reliable system, the Firestreak consistently showed an 85 percent success rate in testing.

A relatively large system, the Firestreak was ten and one-half feet in length and nine inches in diameter, with a wing span of two and one-half feet.

SRAAM Advertisement. (Hawker Siddeley Advertisement)

Hawker Siddeley Red Top

Although basically a modification of the Fire Streak, the Red Top received a new name and significant new capabilities. The system sported increased size control surfaces for improved maneuverability at all altitudes. A hemispherical nose cone covered a new IR guidance system.

The warhead had been relocated further forward in the fuselage next to the fuzing system. The Red Top also carried a more powerful rocket engine and an intercept capability from any direction including a collision course, an improvement over the Firestreak.

A foot longer than the Firestreak, the Red Top possessed a Mach 3 cruising speed and a range of seven miles.

British Aerospace Sky Flash

Working with the U.S. Raytheon company and using the Sparrow AIM-9E airframe, motor, and warhead, the system used a new semi-active homing guidance system. With this system, the Sky Flash had the capability to distinguish among multiple targets and select an individual target to attack. The Firestreak also had an improved capability intercepting targets flying at low altitudes.

The system, which became operational with the RAF in 1980, was popular with foreign buyers, with Oman, Saudi Arabia, and Sweden making purchases. It was carried by the British Toronado and RAF Phantoms. The system was also flown on the U.S. F-16. Thousands of the system were produced.

The system was 12 feet in length with a 3.3 foot wing span and an impressive Mach 4 speed capability with an altitude capability of up to 65,000 feet.

British Aerospace AIM-132 ASRAAM

With a history that started in the 1980s, the Advanced Short-Range Air-to-Air Missile (ASRAAM), which was developed with France, could well be one of the prime Air-to-Air missiles in the next millennium.

The system is produced jointly by BAe Matra Dynamics of France and British Aerospace. Interest in the system has continued to grow internationally, with the Austrailian Air Force selecting the system in 1998 to equip that country's F/A-18. The ASRAAM is also compatible with the Eurofighter 2000, the Harrier GT.5 and GR.7, and later the F-15, F-16, F-18, and F-22.

The ASRAAM uses a solid-fuel motor incorporating significant reductions in detection signature. The motor features an advanced carbon-fiber case. THe ASRAAM has a hypersonic speed capability with a range capability from one thousand meters to about five miles.

The sophistication of the system was at such a high level that it was considered a prime candidate for the USAF AIM-9X program. Confidence in the program was greatly increased in 1999 with the selection of the ASRAAM as the prime armament for the Austrailian Air Force F/A-18s.

Future British Air-to-Air Missiles

BFMAAM (British Future Medium Range AAM)

This future British system will build on AAMRAAM technology, with several international teams bidding for the contract. With the AMRAAM similarity, the new system should be compatible with the same AMRAAM-compatible fighters with the U.S. Raytheon proposal. The European proposal comes from the Matra/BAe team and goes by the program name of Meteor.

Raytheon's proposal used a liquid fuel ramjet designed by Aerospatiale Missiles. The Meteor is assessed to employ a solid fuel, variable ducted ramjet developed by DASM.

The BFMAAM will use a mid-course guidance/active radar seeker guidance system. It will equip initally the Eurofighter 2000 and other aircraft later.

Shorts Star Streak.

Although the British Star Streak is a suface-to-air missile, the system could well be serving in an AAM role in the next millennium. In the late 1990s, the missile, which uses a beam rider guidance system, was being tested to serve as a possible candidate to arm the U.S. AH-64 Apache helicopter.

Chinese Air-to-Air Missiles

Although China might have the largest Army in the world, its equipment is relatively aniquated when compared to the West. During the border war with Vietnam in the late 1970s, the deficiencies of their weapons were well noted. The Chinese have long been interested in the development of Air-To-Air Missiles, but the technology of these systems has also lagged. The country has developed a small number of native systems, but for the most part has depended upon the purchase of foreign systems to fulfill their needs.

CATIC PL-2/2A

The PL-2 was the Chinese version of the Soviet AA-2 Atoll missile and a mainstay during the 1960s, with the system starting production in 1964 and ramping up to series production three years later. It was carried by the Chinese F-6 and A-5 fighters.

An A version of the system would follow to correct problems with the initial version, which was jammable by IR countermeasures. The modified system had a tail chase capability and improved low-altitude capabilities.

The PL-2A had a range of about 11 miles when series production began in 1982, and the warhead was reported to have an effective kill distance of about 11 meters. Over 62,000 of both versions are estimated to have been produced.

CATIC PL-4

The first Chinese medium range system had an external similarity to the U.S. AIM-7 system. Internally, though, it was more like the Soviet Atoll with a semi-active guidance system.

CATIC PL-5B

Significant improvements were the byword in the next PL-5B system, which included an improved seeker, a single-stage solid-propellant motor which provided improved maneuverability capability, and an off-boresight launch capability. The system is very similar to the U.S. Sidewinder, having almost an identical configuration and weight.

Chinese improvements to the design included both IR and RF proximity fusing and two types of warheads, fragmentation and continuous rod. The system also had a Mach 4.5 capability and was operational for a period with the Chinese Air Force.

CATIC PL-5E

Another version of the PL-5 system, the PL-5E, is similar to the Sidewinder. The system has a maximum range of 14km with a maximum lateral acceleration of about 40Gs.

CATIC PL-7

The PL-7 system has great similarities to another Western system, the French Matra R.550 Magic. This program began in the early 1980s and was manufactured by the CATIC organization. The PL-7 is 2.72 meters in length and weighs about 89 kg with a Mach 4.5 capability.

PL-9

The PL-9 was another member of the Chinese AAM family that had another country connection, this time with the Israel-designed Python III. The PL-9 is a licensed Chinese-manufactured version of that system.

PL-10

A new indigenous system, the PL-10 with a semi-active guidance system came on the scene in the mid-1980s. There are indications that the PL-10 was similar to the U.S. AIM-7E, and might have used some of the technology from its HQ-61 surface-to-air system. The effort is thought to have been disontinued in the early 1990s.

PL-11

The newest identified Chinese system, the PL-11, was first noted in 1996. It is a medium range system and uses an active radar system. There are indications that the system might have some close similarities to the Italian Aspide and Russian AA-10 systems.

French Air-to-Air Missiles

Since the 1950s, the French have been involved in the design, development, and deployment of Air-to-Air missiles. A number of them are rated as some of the best in the world. In recent years, the international reputation of the country has been greatly enhanced by a collaboration with the British in the joint development of the ASRAAM missile.

Following is a chronology of the numerous missiles that have been developed by the French aerospace industry.

ASRAAM missile shown on F-16. (British Aerospace Photo)

Matra R.511

Designed in the early 1950s, this early AAM made its first test flight in 1956. The system then went into series production, and during the 1960s served as the armament for the Vautour II.N and Mirage III-C fighters of the French Air Force.

The capabilities of this early system were, as would be expected, somewhat limited, with firing only being available during a pursuit course. The launching altitude range for the system was between altitudes of 10,000 and 59,000 feet.

Matra R.530

The R.530 was the first of two 530-series missiles, assuming an IOC status in 1963 and continuing on operational service into the 1980s. An estimated 4,000+ missiles were built, many of which were adopted by a number of countries, including Argentina, Australia, Columbia, Egypt, Iraq, Jordon, Lebanon, Libya, Morocco, Pakistan, Saudi Arabia, South Africa, Spain, and Venezuela.

The missile was a relatively large system, being almost 11 feet in length, .9 feet in diameter, with a 3.6 foot fin wing span and a weight of about 425 pounds. With either a radar homing or IR guidance system, the R.530 had a range of approximately 11 miles, a top speed of Mach 2.7, an altitude capability of 69.000 feet, and a 59 pound high explosive warhead.

The missile's long cylindrical body was fronted by a blunt nose, and carried large delta wings at mid-body with ailerons on one pair for roll control. In addition, there were also small control fins at the tail for additional maneuverability capabilities.

In operation, the R.530 was normally deployed in pairs for greater versatility. With either type of guidance type, it was necessary for the missile seeker to acquire and lock on to the target before launch could occur. With the IR type, the missile was capable of locking on to the target from either the side or rear, or on aerodynamic hot spots. It was carried on the Mirage III, Mirage F1, and the U.S.-built F-8E Crusader.

Matra Super .530

The Super .530 was the follow-on upgrade to the R.530 system, which doubled the range and seeker acquisition distance of the original system. The design goal of the system was to have a "snap-up" capability from launch to intercept targets flying at up to 30,000 feet. Its maximum speed capability approached Mach 5 with an impressive 25 mile range.

Still a significantly-sized missile at 11.6 feet in length, .9 feet in diameter, and a three-foot fin span, the super .530 weighed in at a hefty 550 pounds.

The system had maneuverability, speed, and range improvements over the initial .530 version. And although it carried the same .530 designation, the Super version was a completely new system with no common parts with the original.

MBB Kormaoran missile under wing of F-104G. (MBB Photo)

Guidance was achieved with a semi-active homing system mated with a target illumination radar in the F version, while the D model used a pulse doppler reader homing. Control of the system with both guidance techniques was accomplished through electrical activation of the tail control fins.

The Super .530 was used on various versions of the Mirage fighter, with Ecuador, Egypt, Iraq, Kuwait, Libya, and Morocco being the using nations. Other applications include the Rafale, F-16, F/A-18, and Tornado.

Matra Magic R.550

The Magic R.550 AAM was designed by Matra as a replacement for earlier versions of the U.S. Sidewinder missile. Emphasis in the design of the new missile was towards improved launch capabilities in dogfight conditions. The missile mounted a pair of fins and was powered by a single- stage solid propellant rocket motor.

The technology of the guidance system was greatly increased with a passive IR seeker cooled by launcher-contained coolant. A mechanical scanner accomplished target acquisition with an onboard servo used for control surface activation.

The major aircraft that have carried the R.550 include the Mirage, F1, Crusader, Jaguar, Sea Harrier, SUper Enendard, Macchi 326K, Hawk, and Alpha Jet.

Matra Magic

The development of the initial Magic model began in the late 1960s and assumed operational status with the French Air Force in 1975.

Matra Magic 2

Then came the Magic 2, which was tested in the early 1980s and quickly assumed operational status in 1985. The system incorporated significant improvements over the initial version with digital signal processing, improved aerodynamics, a refined seeker head, and all-aspect imaging.

The Magic 2 can be operated in a pair of operating modes: the automomous mode, where acquisition is the entire responsibility of the seeker head; and the integrated mode, with the initial target data being provided by the launching aircraft. From an economical consideration, the Magic 2 is completely compatible with the Sidewinder launcher, a system the French have used for a number of years.

Matra MICA

The MICA is the next in the long line of Matra-designed Air-to-Air missile system. The MICA was designed for the Mirage 2000 and the future Rafale aircraft. The goal of the MICA is a system capable of both short and medium range intercepts.

Two versions of the Mica, the -IR and -EM, are being developed. Both version are ten feet in length, 6.3 inches in diameter, have a high-explosive warhead, and a Mach 3 capability. The maximum range is about 30 nautical miles with minimum range intercepts possible at .27 nautical miles.

The advanced French system uses an imaging infrared system in addition to a semi-active homing arrangement. With a range of about 50 km, the MICA was being exported to a number of countries in the late 1990s. As of this date, though, the system had not been cleared for the U.S. F-16, greatly decreasing its export possibilities.

However, in the late 1990s the United Arab Emirates purchased both versions of the MICA. The systems would arm the new Mirage 2000-9, which was also purchased. The transaction was a significant first for exporting the MICA system.

Matra .550 Magic missile began replacing early Sidewinder missiles in the mid-1970s. (Matra Photo)

A Mirage 2000-5 fighter carries a full weapons load, including the Mica missile. (Matra Photo)

Future French Air-to-Air Missiles

Meteor Missile

Matra, during the late 1990s, was also involved in developing a new AAM for the Eurofighter 2000. The goal of the new design, coined the Meteor Missile, was to achieve superiority over current and future threats to the year 2030. The Meteor was also a candidate for the British FMRAAM program, and was also a part of the BVRAAM (Beyond Visual Range Air-to-Air Missile) program.

The new system was being designed by an international team with the goal of being compatible with a majority of U.S. and European aircraft. The ambitious design goals called for the system to be three times more effective than the AMRAAM.

In an interesting deviation from the norm, the Meteor was designed to carry a ramjet propulsion system, which enabled it to meet its kinetic, range, and end game requirements.

Matra MICA Mark 3

Matra in 1998 was also considering the design of a Super Matra, to be called the MICA Mark 3, but no final decision had been made at presstime.

ASRAAM AAM

The Advanced Short Range Air-to-Air missile (ASRAAM), a joint British/French development with MICA overtones, is covered in the British section of this chapter.

Joint British/French VSRAAM

In the late 1990s, a three-year study was initiated for a Very Short Range Air-to-Air Missile (VSRAAM), an AAM that would provide an active defense against air-launched and surface-launched missile threats. Officials indicated that such a system could be deployed by 2015.

The air-launched weapon suite of the Swedish Gripen is laid out. It includes the MICA AAM. (Matra Photo)

German Air-to-Air Missiles

Eurofighter Air-to-Air Missile (EURAAM)

In order to cope with future threats and to create a European alternative for a medium range Air-to-Air missile suited primarilly to arm the Eurofighter, the German aerospace industry considered the development of a new AAM.

To that end, Daimler-Benz Aerospace laid out a new Air-to-Air missile program based on the A3M missile, a concept designed during a company and German government funded technology study and demonstration program.

Great Britain, Italy, Spain, and Sweden were also initially in volved with the system.

The late 1990s study was able to fulfill the UK Requirement SR(A)1239 and was a system that was totally compatible with the U.S. AMRAAM system. The propulsion system planned was a ramjet solid propellant system and was estimated to have capabilities far surpassing current capabilities.

The missile was programmed to carry a high-performance radar seeker which possessed high ECM resistance. In terms of operation, the missile was to fly toward the target according to a multiphase guidance concept in order to attain a long-enough range and to achieve high accuracy.

A mid-course phase took place with inertial command guidance to bridge longer distances, backed up by target position and speed updates. Missile velocity was to be controlled and would be adjusted to the specific engagement conditions.

The missile was to be flown mainly in the bank-to-turn mode, and only used at the end of the terminal gudiance phase and shortly before intercept guidance is effected in the skid-to-turn mode. A radar proximity fuse initiates detonation of the preformed fragmentation warhead within the lethal distance from the target.

BGT Armiger AAM

In the late 1990s, the Bodenseewek Geratetechnik (BGT) Company of Germany was developing the Armiger, an AAM system employing a Mach 3 ramrocket propulsion system which was thought to provide increased range and shorten time to target.

The proposed missile would feature four inlets at the center body location. After being boosted to the required ramjet operating speed by the solid booster, the ramjet sustainer would take over for the terminal portion of the flight to target.

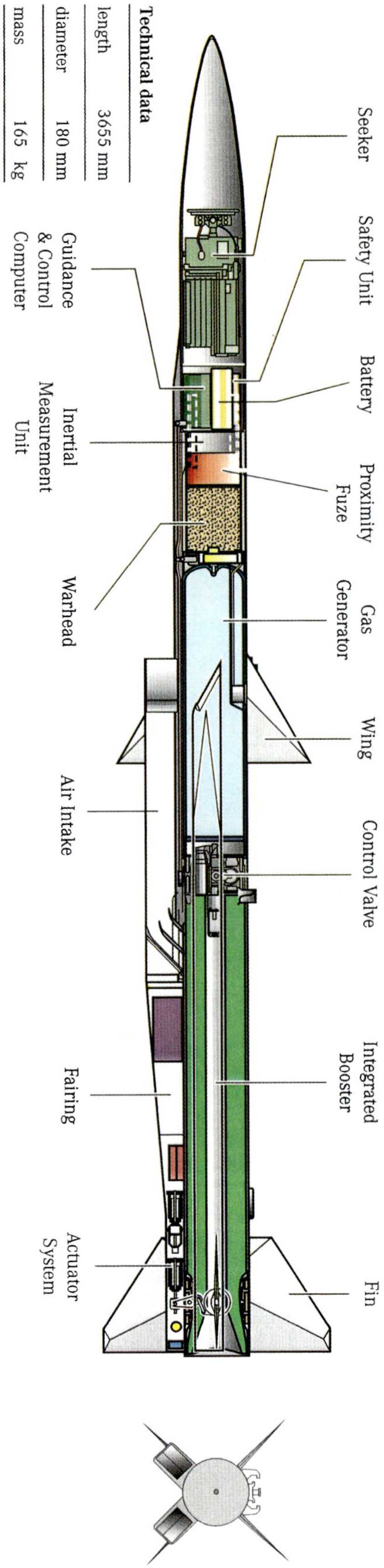

Internal layout of the EURAAM AAM. (Daimler-Benz Drawing)

A EURAAM is carried aboard a Eurofighter. (Daimler-Benz Photo)

Daimler-Benz Medium Range Air-to-Air Missile

During the mid-1990s, Daimler-Benz conducted a technology study on a Medium Range Air-to-Air Missile system. The program was financed by the German government.

The pure technology portion of the program was conducted during 1995 and 1996, with the demonstrator program being carried out through the end of 1997.

Propulsion was unique, with a boron-ducted vaiable flow ducted rocket motor. The design also considered the use of a sophisticated electro mechanical solution with respect to high dynamics under heavy loads, with heat protection accomplished by a combination of thermal barrier and heat sink mass.

Indian Air-to-Air Missiles

Astra AAM

The Astra missile began development during the mid-1980s, with the first official announcement of its existence coming in 1989. The missile was designed to arm the Light Combat Aircraft (LCA). It is suspected that technology from a number of other foreign sources would be incorporated into the system.

Little information is available on the system, but it is assessed that it will use a solid-propellant rocket motor with possibly inertial guidance and active homing.

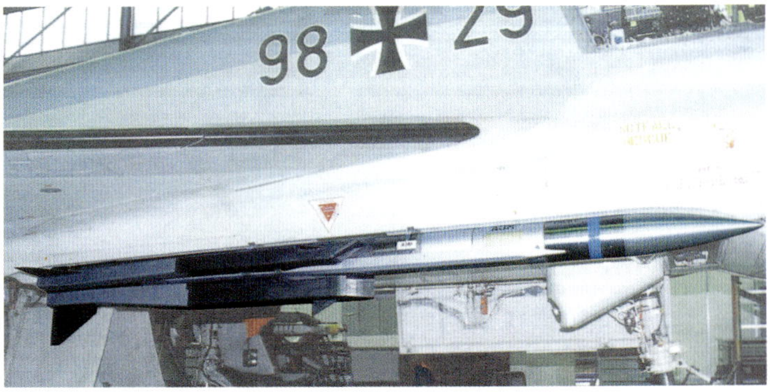

A EURAAM is mounted on a German Air Force F-104 fighter. (Daimler-Benz Photo)

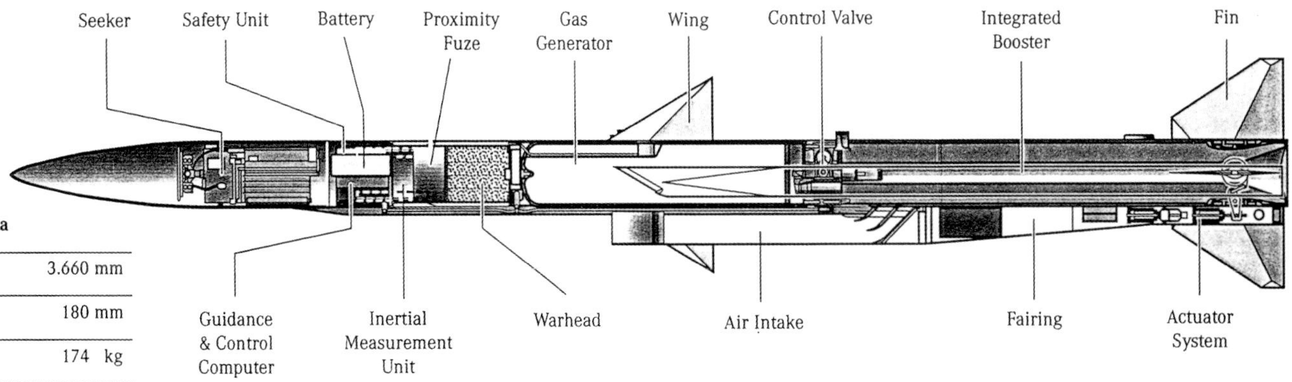

Technical Data

length	3.660 mm
diameter	180 mm
total mass	174 kg

Cutaway drawing of Daimler-Benz MRAAM (Daimler Benz Drawing)

Shafrir under the wing of a Kfir fighter. (Israir Photo)

Israeli Air-to-Air Missiles

With a strong aerospace technical base, Israel has developed a number of outstanding Air-to-Air missiles since the 1960s. As the next century approaches, the country is continuing its research with new systems continuing to appear.

Rafael Shafrir II

One of the first Israeli AAMs was the Shafrir, whose development began in the mid-1960s and which assumed operational status in 1969. The infrared guided missile has seen considerable combat experience through the years. An extremely simple system to operate, pilot training consisted of only one day of instruction and no test firings. The system in its early days exhibited a 50 percent kill capability. Its attractiveness came from its minimal unit cost of only $99,000 per copy.

A relatively small weapon, the Shafrir was carried under the wing in a special adapter and was powered by a solid propellant rocket motor. When the target was detected, an audio signal was heard and a light illuminated on the panel, indicating to the pilot to press the firing button.

The system was exported to a number of countries, including Argentina, Chile, Ecuador, and Taiwan, among others, but production ceased in the late 1990s.

Rafael Python III

The Python III Air-to-Air Missile was the follow-on design to the Shafrir and had three times the range of the Shafrir. The system is almost ten feet in length with a diameter of only about six inches. Weighing about 265 pounds at launch, the Python III carries a warhead weighing about 25 pounds. The guidance is provided by a passive infrared homing system.

The Python III had combat initiation with the Israeli Air Force in a confrontation with Syria in 1982. With Mach 3.5 speed and 40g turn capabilities, the Python III quickly showed its superiority with a reported 50 kills.

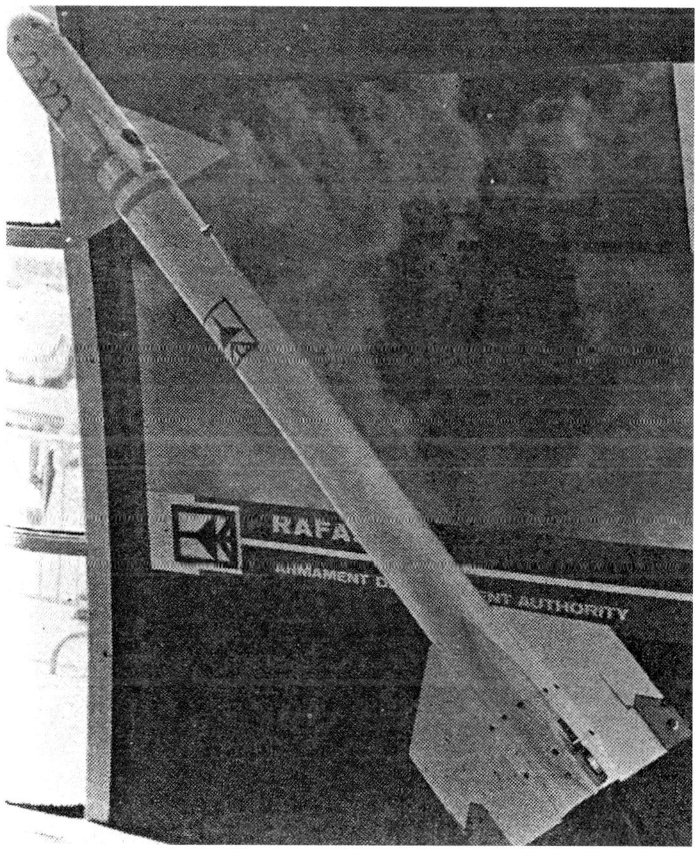

The Shafrir missile bears a marked similarity to the U.S. Sidewinder. (Israir Photo)

Rafael Python IV

The Python IV was the newest Israeli AAM in the late 1990s. The system was introduced in 1993, but very little information has been released about the system. The system fulfilled the Israeli Air Force's need for a short range system with greater off-boresight capability, reported at 60 degrees.

The missile is six inches in diameter and has four rear wings with forward control surfaces. Slaved to a helmet-mounted display/targeting system, the Python IV incorporates IR and rocket-motor smoke reduction.

The system was in the news in the late 1990s as it, along with the U.S. AMRAAM, will be the prime armament for the new F-15I, which Israel just started receiving. It is also qualified for the F-4, F-5, F-16, F/A-18, and the French Mirage 2000.

In 1998, an agreement between Rafael and the U.S. Lockheed Martin Company was signed to market and improve existing systems.

The Python IV was the first improvement effort, with production taking place in both countries. At press time, there were indications the research could carry beyond the Python IV.

Rafael Python V

Israel is also working on a missile coined the Python V. There was discussion that the missile might use the seeker planned for the AIM-9R Sidewinder. The status of the program is unclear at press time.

Rafael MRAAM/Possible F-16 Compatible Missile

It has also been reported that a Medium Range Air-to-Air Missile (MRAAM) has been under development during the 1990s, and that an MRAAM-type missile has been developed for the IAF fleet of F-16s. Could both reports be of the same missile?

Italian Air-to-Air Missiles

The Italian aerospace industry has produced a number of anti-aircraft missiles for its own military, along with many for export.

Salenia Aspide

The 1980s-vintage Aspide AAM was a medium-range, multi-role missile based on the U.S. Sparrow. The missile was adapted to a number of ground and sea-launched air defense missions. The system was motivated by the fact that the U.S. declined to export the AIM-7M system at the time.

Assuming operational status in 1987, the Aspide had a Mach 2 speed capability and carried a 73 pound warhead. The system was reported to be very effective at both high and low altitudes and in a heavy ECM environment. It regularly arms the F-104Ss of the Italian Air Force.

In the late 1990s, the Mk2 version of the Aspide with an improved seeker, has been developed.

Japanese Air-to-Air Missiles

Although they started late in the Air-to-Air Missile arena, the Japanese have come on in the 1990s with several native-developed systems. Japan also produced the AIM-7F Sparrow and AIM-9L Sidewinder under U.S. license.

Mitsubishi AAM-3 (Type 90)

Design and development work on this highly-maneuverable AAM began in the mid-1980s. The goal of the program was to replace the AIM-9L. Testing was completed in 1989, and it entered operational service in 1992, being carried by the F-15J and F-4J fighters.

The AAM-3 is almost identical in size to the Sidewinder, being 9.5 feet in length, two feet maximum fin span, and 200 pounds in weight. The propulsion is provided by a solid propellant rocket engine, and guidance is produced by an IR system. The reported range is 2.7 miles. At the turn of the century, the AAM-3 was still in production.

Mitsubishi AAM-4

This system began full-scale development in the early 1990s to replace the AIM-7 Sparrow. Little is known about the status of this development, except that it is expected to enter service in 2000. It is estimated, though, to carry a dual-capable seeker head capable of providing semi- and fully-active modes.

The Japanese Navy reportedly wants to convert the system into a short-range, sea-launched version as the replacement for the sea-launched version of the Sparrow.

Russian Air-to-Air Missiles

Through the Cold War, the then-Soviet Union matched the West AAM for AAM, in both quality and quantity. At the end of the 20th century, the greatly-reduced military complex still is producing high-tech Air-to-Air missiles with significant capabilities. Following is a discussion of the capabilities and characteristics of the Russian family of AAMs.

AA-1 (R-13 Soviet Name)(NATO name ALKALI)

The first Russian missile, the AA-1, was designed in the 1950s and became the standard armament for the MiG-19 and the Su-9 fighters. The system used a solid-propellant propulsion system with large rear and smaller front fins. Control surfaces on the rear of the rear fins enabled the missile to maneuver on commands from its semi-active homing guidance system.

AA-2 (R-13A Soviet name) (NATO name ATOLL)

The second Russian AAM appears to have been the result of exact reverse engineering of the U.S. Sidewinder missile. The AA-2 is almost identical in size, shape, guidance system, and solid propellant propulsion system. It was long the standard armament for the MiG-21 fighter and saw considerable action in Viet Nam.

The AA-2 has the characteristic long cylindrical body with control surfaces far forward and cruciform tail surfaces. The forward control surfaces have a leading edge angle of about 60 degrees and ten degrees on the trailing edge, and have a movement capability of about 30 degrees.

The success of the AA-2 resulted in a number of different versions being built. The AA-2-2 system reportedly changed the guidance technique to pure radar guidance, a system which would be called the Advanced Atoll.

The system was recognizable from its longer nose and canard control surfaces. The system was also built under license in India, and also served as the basis for the Chinese PL-2 AAM.

AA-3 (NATO name ANAB)

The AA-3 was first seen at the Soviet Aviation Day Display in 1961 mounted on a Yak-28P fighter. The widely-used system was used on the Su-9, Su-11, and Su-15 fighters. The system had a cylindrical body with a pointed nose (for the semi-active radar homing type) or a blunt nose (for the IR guidance system), and a taper on the tail. There was a delta wing assembly at mid-fuselage with forward canard controls.

Propulsion was accomplished by a solid propellant motor. The warhead was a high explosive blast type which was detonated by a proximity fuze.

Reportedly, there were attempts to upgrade the capabilities of the system, but some problems were encountered. It is assessed that the system was probably replaced by the later AA-7 Apex system. Probably several thousand of the missiles were produced. Besides the USSR, the system was used by Bulgaria, Czeckoslovakia, East Germany, Hungary, Poland, and Romania.

AA-5 (NATO name ASH)

The size of the AA-5 was in line with the large size of the Tu-28, being about 17 feet in length with a 4.3 foot wing span. Like the AA-3, there were two different guidance versions of the system; the blunt nose indicated the IR version, and the pointed nose system was the SAR version.

It is assessed that probably several thousand of these missiles were produced with IOC occurring in the early 1960s. The Tu-28s normally carried four AA-5s on patrol missions. Also, early versions of the potent MiG-25 were noted carrying the Ash missile. Using nations included Bulgaria, Czechoslovikia, East Germany, Poland, and Romania.

An A-5 under the wing of a Libyan Air Force MiG-25. (U.S. Navy Photo)

AA-6 (R-40/46 Soviet name, NATO name ACRID)

The AA-6 was identified in the mid-1970s as the standard air-to-air missile for the MiG-25 and operated with a fire control radar system known as the Fox Fire. The system is also carried by the Su-15 D and E models, along with the MiG-31 and Su-35/37 models.

The system is similar to the AA-3, but considerably larger. It is suspected that there are both an IR and a radar-homing version of the system. The system is thought to be about 20 feet long with a range of almost 40 miles. The AA-6 is larger than the U.S. AIM-54 Phoenix system, but is not nearly as effective at the longer ranges of the U.S. system.

The system was used on MiG-25s by a number of Soviet equipment-buying countries like Algeria, Iraq, Libya, and Syria.

AA-7 (NATO name APEX)

The highly effective MiG-23 carried the AA-7 as its standard missile, making the combination a signficant air-to-air challenge. The system has also been used on the MiG-25E fighter. The AA-7 has been compared to the U.S. Sparrow missile, and like many other Russian AAMs, comes in both an IR and a radar homing version. The AA-7 is thought to have replaced the AA-3 system.

Capabilities are significant, with the capability to be launched from 150-to-80,000 feet. It is estimated to be most effective against low-to-medium altitude targets.

The airframe for the AA-7 sports a long cylindrical body with a blunt nosecone. Large delta wings are at mid-body, while smaller canards are located on the guidance section. Power comes from a two-stage solid motor. The warhead weighs in at about 88 pounds and uses proximity fuzing.

AA-8 (Soviet name R-60, NATO name APHID)

The AA-8 is the second type of Russian AAM that serves as a standard missile for the MiG-23, along with also being carried by the MiG-31, Su-24M, Su-24M, Su-27, Su-35, and Su-37. The system is a short-range solid-propellant missile that uses IR guidance and radar homing systems, in two different versions.

The AA-8 is thought to have replaced the AA-2 system and is assessed as having much better maneuverability. It is possesses extremely high G maneuverability even at high altitudes. The system was designed in the late 1960s as a strictly pursuit missile.

The system was also used in an operational environment by the South African Air Force during operations over Algeria. Users of the AA-8 have included Angola, Cuba, Czchoslovakia, East Germany, India, Iraq, Libya, and Syria.

The AA-8 has a fuselage length of about seven feet with a maximum fin span of 1.3 feet and weighs about 121 pounds. It has a cylindrical body with a blunt nose. There are large delta wings aft and in-line cropped-delta canards aft of the ogive. In addition, there are four small blades located on the nose.

The AA-8 is thought to have become operational in the mid-1970s, but exact production numbers and dates are unkown.

AA-9 (Soviet name R-33, NATO name AMOS)

Like every other type of Soviet AAM, there seems to be a western equivalent, and that is the same with the AA-9, which seems to bear close similarities to the U.S. AIM-7 and the British Sky Flash. It uses an inertial autopilot guidance system with seimiactive radar terminal homing. A large missile, the AA-9 has a length of 13 feet, seven inches, and a body diameter of one foot, three inches. The 1,080 pound missile carries a 104 pound blast fragmentation warhead.

The system seems to be mostly associated with the MiG-31 fighter interceptor. It is thought to have a maximum range capability of about 45 miles, and possibly twice that figure. The AA-9 has demonstrated amazing capabilities at low altitudes, reportedly intercepting a drone target at 150 feet. It has also been reported that the AA-9 was tested against simulated cruise missile targets.

AA-10 (Soviet name R-27, NATO name ALAMO)
One of the later AAM versions, the AA-10 has been noted on the MiG-29, Su-27, Su-25, and Su-27 fighters. The system has a range capability of between 50-to-75 miles depending on the particular version, of which there are many.

The versions are combinations of various rocket motor/guidance system combinations.

The AA-10 A version uses a semi-active homing, short rocket motor, the B version uses IR guidance/short rocket motor, and the C version is a semi-active guidance/long rocket motor combination, while the D version is an IR guidance/long rocket motor grouping. The versions vary between 11 and 15 feet in length, and between 440 and 529 pounds in weight.

AA-11 (Soviet designation, R-27, NATO name ARCHER)
A true dogfight missile, this modern system is carried by the MiG-29 and Su-27 fighters. Four of the missiles are usually carried by the latter fighter. Powered by an advanced solid propellant motor, the system uses canard fins and fins in the rocket exhaust for thrust vectoring to accomplish its great maneuverability capabilities. The system is thought to weigh about 275 pounds, is about ten feet long, and only seven inches in diameter. It uses an active radar proximity fuze on its 33 pound warhead.

The AA-11 is thought to be Russia's most advanced heat-seeking missile, and could well be close to the performance of the U.S. Sidewinder. It is thought to have a combat range of about 18 miles. Two versions have been identified, the M1 version with a 45 degree off-bore sight capability, and the M2 version with 15 degrees greater capability.

AA-12 (Soviet Designator, R-77, NATO name ADDER)
The R-77 AA-12 ADDER system is thought to be near operational service, having entered production in the late 1990s. This system is thought to use active radar homing guidance, possesses extremely high performance, and is assessed to be effective against high-G capable fighters. The MiG-29M, MiG-35, Su-30, and Su-34 fighters are assessed as being capable of carrying the AA-12.

The missile is almost 12 feet in length and weighs 496 pounds in the heaviest version. It carries a 48 pound continuous rod fragmentation warhead.

AA-13 (Soviet designator, K-13)
The AA-13 is a greatly advanced version of the AA-9 and has been sighted on the MiG-31M. The new system is powered by a solid propellant propulsion system and an inertial, with an active terminal homing guidance system, and a 132 pound fragmentation HE warhead. Its range is an impressive 93 miles.

The AA-13 carries folding fins which could allow for internal storage in aircraft fuselages.

K-37/R-37 AAM System
This system is presumed to be the replacement for the AA-9 Amos. Little is known about the advanced system, except that it will have longer range and a more powerful warhead than the system it is replacing. It also uses an advanced guidance system consisting of inertial navigation and mid-course updates by means of a data link.

Future Soviet AAMs
There is a continuation of Russian AAM designs which will undoubtedly appear in the next decade.

South African Air-to-Air Missiles

South Africa has quietly been developing a family of Air-to-Air Missiles over the past three decades. There continues to be interest in future developments.

Kentron V3 Series of Air-to-Air Missiles
There have been a number of versions of the V3 series of missiles, including the V3A, V3B, V3C, and the latest, the V3U.

Kentron V3A/V3B Kukri AAMs
The initial A and B versions bear a close resemblance to the French R.550 Magic system. These systems are no longer in production, but remain in service with the South African Air Force in the late 1990s, being carrried on Mirage fighters.

These first versions of the V3 carry a high explosive fragmentation warhead, are powered by a double-base, solid-fuel rocket motor, and have a twist/steer guidance system. The concept uses fixed forward fins with the forward cruciform fins being movable, along with the free-swiveling rear fins and complete aft section of the missile. The missiles are capable of completing turns up to 7G.

Deliveries of the V3A began in 1975, which was the same time that the V3B design work began. The A and B modifications have been exported under the Kukra name, with Jordan being an identified customer.

Amscor V3C/V3U Darter
In 1988, South Africa displayed for the first time the next modification of the V3, the V3C, which was considerably different from the previous systems. The maneuverability had been greatly increased, along with advances in seeker technology. The V3C is identical to the A/B version in size.

The V3C began series production in 1987 and is estimated to have assumed operational status the following year.

The evolution of the system continued in 1993, when an even more improved version of the V3—the U-Darter—entered series production. The U version has a longer range with a more powerful rocket engine and a dual-band IR seeker head. There is also a 50 degree off-boresight capability, and it can also sustain higher G loads.

Kentron V3S Super Darter
(Also called A-Darter or Agile Darter)

This highly maneuverable system served as an interim step in the late 1990s for the next higher-technology system which was under development during the period. The V3S is unique in that it is the first in the long line of V3 systems to vary from the long-standing airframe.

The exact status of the development program in the late 1990s was uncertain, having been in progress for about five years. If a go-ahead for full-scale development were to have been given in 1998 or 1999, it is presumed that the system could be ready for series production between 2002 and 2005.

The A-Darter is carries a new high-speed computer and seeker, the latter based on imaging infrared technology with a greater look angle than the earlier members of the V3 family. For greater maneuverability, the system will support a new thrust vector control system with tail control.

Kentron V4 R-Darter

The change in V designation points to the fact that this Darter is a different animal than the V3 series. The R in the name indicates Radar-Guided, which is quite a departure from the IR guidance of all the previous members of the V3 family. It has been reported that the system might be a lower-cost, simpler system, with development being completed early in the next millenium.

Kentron Medium Range AAM

During the early 1990s it was reported that South Africa was involved with the development of a Medium Range AAM. There is the possibility that this work also involved the Israelis. The exact status of this system is unknown, although there are indications that the effort might have been terminated.

Future South African Air-to-Air Missiles

Long Range Air-to-Air Missile (LRAAM)

Another interesting South African proposal for the next century is the LRAAM system. The missile is based on the U-Darter airframe, but little beyond that is known about the development effort.

Swedish Air-to-Air Missiles

There have been a number of Swedish Air-to-Air missles that were produced under license to the developing countries. The systems included the J and L versions of the U.S. Sidewinder and the British Sky Flash. The systems were identified as the RB-24, RB-74, and RB-71, respectively. The first two systems would serve as prime armaments for the Saab AJ37 Viggen fighter.

It has also been reported that the Swedes studied a pair of Air-to-Air systems. The SAIR (Swedish Advanced Infrared) system, a Saab design effort, was to be a short-range attack missile for the Gripen JAS39 fighter. There was also consideration for converting the RBS.70 Surface-to-Air missile into an anti-aircraft system for helicopter armament.

Taiwanese Air-to-Air Missiles

Taiwan has been involved with the design and production of two native designs, although the first system (the Sky Sword 1) is a close copy of the AIM-9M Sidewinder.

CSIST Sky Sword 1

With an IR guidance system, the Sky Sword 1 is 9.4 feet in length, has a two-foot maximum fin span, and weighs about 198 pounds. Its range is reported at 2.6 nautical miles.

There are also reports that there may be an upgrade to the Sky Sword 1 with a Sky Sword III version.

CSIST Sky Sword 2

The second generation Sky Sword 2 is considerably larger than the initial version with a length of 11.8 feet, maximum fin span of 2.5 feet, and a launch weight of 264 pounds. Instead of an IR guidance system, this system is reported to be using an active radar guidance system.

The system, also known as the Tien Chien I, is assessed to be in the same category as the U.S. AIM-7.

Significant Canceled Foreign Air-to-Air Missile Systems

Even with the multitude of AAMs that have been developed through the years all over the world, there have been nearly as many systems that were discarded in various portions of their programs. Some did not make it beyond the study stage, while others would have prototypes built and tested. Following is a discussion of the more important systems, discussed by their country of development.

French Air-to-Air Missiles

The French Matra company, which had momentum from its successful design and development of the 530 and Magic AAMs, looked at the development of the advanced MICA system during the 1980s.

Advanced MICA AAM

The MICA was planned to be the prime air-to-air weapon for the French Air Force from the mid-1990s and into the next century. The system was designed with the goal of countering the expected threats of the time period.

The versatile system had several modes of operation. For close in combat, the system was locked onto its target before it was launched. For longer engagements, the MICA's inertial guidance system was used until the final seconds when the onboard seeker would come into play.

Flight testing began in 1989, with operational service planned for the mid-1990s. It would not happen, though, as the system would be canceled, possibly a casualty of the newer ASRAAM.

Italian Air-to-Air Missiles

Selenia Idra AAM

A possible AMRAAM-class missile, called the Selenia Idra, was under development in the 1990s which used Aspide technology, along with a fire-and-forget AMRAAM-style guidance system. The program, however, was reported to have been combined with the French/Italian ASTER program.

South African Air-to-Air Missiles

Medium Range AAM

As mentioned in the South African section, there was a medium range AAM program that was reportedly co-designed with Israel. It is estimated that the system was terminated before full-scale development.

4
U.S. Air-to-Ground Weapons

It is said in many Air Force squadron ready rooms that fighter pilots make movies, but bomber pilots make history. But there is some truth to the statement, despite what the fighter pilot community might want to believe. No war was won, no conflict resolved by shooting down enemy airplanes. It was when air power destroyed targets of value that the war/conflict was resolved. This truth appears from the time of the first aerial attack to the recent Kosovo crisis. Not until bridges are falling into rivers, oil refineries going up in flames, or a country's entire power grid getting turned off did the conflict come to an end.

But in what is perhaps one of the biggest shifts in air power thinking, fighter planes are bombers and bombers are also fighters. The F-117 does not carry a gun, nor does it carry air-to-air missiles, only bombs. The F-15E has a superb air-to-ground capability in addition to its air-to-air mission. F-16s are dropping laser guided bombs, taking out tanks with Maverick anti tank missiles, and taking on surface-to-air missiles in the Wild Weasel role. There is even talk of an F-22 air to ground mission.

Though we probably will not see bombers in a dogfight any time soon, they are carrying brilliant weapons that will search and destroy enemy targets; GPS guided bombs and air launched cruise missiles, able to target tactical targets, not just the strategic ones.

Many aviation historians point to January 1912 as the starting point of air-to-ground operations. An Italian Air Force lieutenant, based in Tripoli, attacked a Bedouin enclave with four small bombs that he tossed out of the cockpit of his biplane. Crude, but effective.

The basic principles of air-to-ground have not changed much since then, only the technology and the capability. But the basic premise of putting an aircraft-delivered explosive device on a ground target remains the same.

In World War II, groups of 200 or more bombers would launch to strike a single target. Often their efforts were not totally successful, and the target would have to be struck on numerous occasions in order to destroy it. Beginning in the Gulf War, and again in Kosovo, a single aircraft would strike a target, destroying it nearly every time. And, for the first time, a single aircraft could attack multiple targets on a single pass—the B-2 armed with JDAMs.

Another vast change that has come to pass is the drastic drop in planes being shot down. Even with the ground defenses getting better all the time, the loss rate continues to drop.

In WWII, bombing runs against some of the heavily defended German sites resulted in loss rates of 20 to 35 percent—upwards of 600 Americans killed or taken prisoner on a single mission! That number dropped significantly in the Gulf War, and dropped again during the 78-day Kosovo air war to two aircraft downed and none killed or taken prisoner.

Air power, especially air-to-ground power, can be directed at virtually anything. From bombing fielded armies to attacking deeply

F-16, F-22, and F-117(L-R), all designated as fighters, but all have (or will have) a significant air-to-ground mission for the military services that fly them.

An AGM-88 high-speed anti-radiation missile (HARM) on an F-16CF. (USAF Photo)

An Airman loads GBU-12 gravity bombs into a B-1 bomber at RAF Fairford during Operation Allied Force. (USAF Photo)

A GBU-31/32 Joint Direct Attack Munition is readied to be loaded on an F-16. (USAF Photo)

buried and protected bunkers, nothing is out of reach of a capable air force. Though this book will not get into air power theory (there are enough of those type of books out there already), one can say that air power, though capable of striking nearly anything, should not. The Geneva Convention and the Laws of War prohibit attacking many categories of potential targets. Churches, cemeteries, and factories not producing war materiel are just a few examples.

The weapons used today are as varied as the types of aircraft that carry them. From nuclear bombs to GPS-guided, unpowered bombs, to TV-guided missiles on bombers, fighters, and helicopters attacking fielded forces, deeply buried bunkers, tanks, or ships, the goal remains the same—put metal on target.

But what is a target? One of the biggest revolutions in applying air-to-ground power is exactly this—what is the target? In WWII, it was pretty easy. If you wanted to stop the trains from running, you blew up the entire rail marshalling yards. Nowadays, putting a single precision bomb into a rail yard control room can have the same effect—the trains do not run, but you have not destroyed the infrastructure. The same is true for power plants, TV stations, etc. A single missile or bomb in a critical area will shut down the whole system, but the system is not destroyed. The big advantage is that once you win the war (an assumption one makes, because if one were going to lose, one probably would want to destroy the entire system), you can quickly fix or replace that key node, and have full access to the system. Just think of what a difference that would have made in Germany in the late 1940s, or Korea in the late 1950s. This revolution in targeting is a huge change in the way the world looks at successful bombing attacks.

No other category of weapons is experiencing such rapid change as air to ground missiles. Every day there is news about a new, improved guidance system that will put a missile first through a window, then a doorway, and now it seems like the missile will be

With no guns or air-to-air missiles, the Stealth plane's main mission is carrying air-to-ground weapons. (USAF Photo)

A pair of A-10 Warthogs head for Kosovo loaded with AMI-9s, AGM-65 Mavericks, and GBU-12s. (USAF Photo)

Even though the B-52 is many decades old, it has been used as a platform for launching air-to-ground weapons. (USAF Photo)

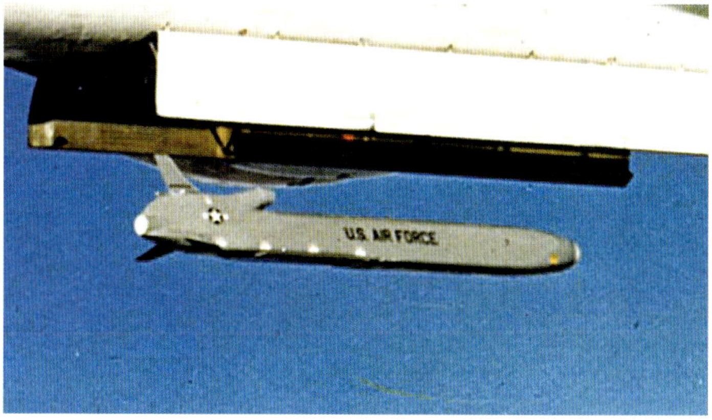

AN AGM-86 is launched from a B-52 during a series of flight tests in the mid-1980s. (USAF Photo)

able to attack a specific brick in the building. One major lesson learned from Operation DESERT STORM was the critical need for a stand off capability. The British took horrendous losses in their Tornado fleet trying to overfly Iraqi airfields to drop their JP233 runway cratering bombs.

Probably the most important technological innovation is the growing use of GPS guidance systems. Not only is the GPS showing up in new weapons, such as the Joint Directed Attack Munition (JDAM), but it is showing up strapped onto regular iron bombs, making them not only smart, but near brilliant bombs. These strap on kits are cheap and fit nearly every bomb in the U.S. inventory.

During Operation DESERT STORM, the U.S. air forces (AF, Navy, and Marine Corps) dropped 210,004 unguided bombs; 9,342 laser guided bombs; 2,039 anti radiation missiles; 5,488 guided air to surface missiles; and 333 cruise missiles. Less than 10 years later, in Kosovo, the numbers were 15,445 unguided bombs and 7,417 precision weapons (laser, TV, or GPS guided).

The air-to-ground category is big and covers a wide assortment—ground attack missiles, gravity bombs, anti-tank missiles,

An F-16 Fighting Falcon from the 416th Flight Test Squadron Edwards Air Force Base, California, fires an AGM-88 HARM missile during testing. (USAF Photo)

cruise missiles, and anti-ship missiles. Gravity bombs, though not launched in the traditional sense of igniting a motor, are so accurate and deadly, and can fly up to 15 miles from the drop point to strike their target, that they MUST be included—for all intents and purposes, they are launched in an unpowered mode.

This chapter is not intended to be the complete guide to every weapon ever developed. If it were, the chapter would run hundreds of pages, but it will look at the most significant programs that made it to operational status, and some programs, though not operational, that had or will have a major impact on air-to-surface weaponry.

The chapter is divided into three subchapters; the first section looks at past and current air launched powered weapons, the second section looks at air launched gravity weapons, and the third section looks to the future.

Past and Current Air Launched Powered Weapons

AGM-12 Bullpup

Manufacturer: Martin Marietta
Native Users: U.S. Navy
Foreign Users: Argentina, Australia, Denmark, Greece, Norway, Turkey, and United Kingdom
Variants: AGM 12A/B/C/D/E
Launching Aircraft: A-4, A-5, A-6, A-7, F-4, F-8, F-100, F-105, P-3, Sea Vixen, and Buccaneer
Operating Period: Bullpup A entered U.S. service in 1959 and remained in production for nearly 20 years. Bullpup B followed four years later.
Mission: Short range, manually guided or automatic IR
The Story: A little confusing, as their were two Bullpups, the -A and -B versions. The smaller A version was initially built around a 250 pound warhead, which was later upgraded to the -E version and carried a fragmentation warhead. The Bullpup B, a much larger missile (about three times the weight, though only 25 percent longer), was called the AGM-12C and D versions. These versions were built around a 1,000 lb warhead. Some -D versions carried a 15 Kiloton nuclear warhead. Neither version serves any longer in U.S. service,

though a number are still in use in several foreign countries. Bullpup A had a 5 mile range, and the B through D versions had a 6 mile range (pretty short for a nuke carrying missile!).

AGM-28 Hound Dog
Manufacturer: North American Aviation
Native Users: USAF
Foreign Users: none
Variants: AGM-28A/B
Launching Aircraft: B-52
Operating Period: 1969-1976
Mission: Air-to-surface nuclear cruise missile
The Story: The Hound Dog was designed to extend the capability of the manned bomber by delivering a nuclear warhead several hundred miles ahead of the launching aircraft, blasting a hole in the enemy defense and allowing the bombers to get through to attack key targets deep in enemy territory. The Hound Dog was a supersonic missile (Mach 2) that could cruise at altitudes above 50,000 feet for distances up to 700 miles. It was a big missile, about 42.5 feet long and weighing 10,000 lbs. For guidance, it used a preprogrammed INS system.

AGM-45 Shrike
Manufacturer: Texas Instruments and Unisys
Native Users: USAF, USN
Foreign Users: Belgium, Denmark, Greece, Israel, Malaysia, Norway, Pakistan, Saudi Arabia, Singapore, Turkey, UK, and Venezuela.
Variants: AGM-45A and -45B
Launching Aircraft: F-105, F-4, F-16, A-4, A-6, A-7, and Kfir
Operating Period: Entered operational service in 1965, some still in U.S. service; in wide use by foreign countries
Mission: Medium range, anti-radar air-to-surface missile
The Story: The first anti-radar missile, developed by the U.S. Navy in the early 1960s using AIM-7 parts. It has seen action in Vietnam, the Middle East, Libya, and Iraq. The Shrike had a range of about eight to ten miles, and incorporated a passive seeker. The 125 pound fragmentation warhead would shred any radar dishes that it attacked—a soft kill, as it did not attack the soldiers manning the radar site, only the radar emitting part of the site. Newer weapons are being developed to provide hard kills on radar sites, closing them down permanently. During the Gulf War, 78 Shrikes were launched by U.S. air forces.

AGM-53 Condor
Manufacturer: North American Aviation
Native Users: U.S. Navy
Foreign Users: none
Variants: unknown
Launching Aircraft: A-6, F-4, F-111
Operating Period: late '70s and 80s
Mission: air-to-surface attack
The Story: It is hard to find much information on the Condor, especially since it was developed by the Navy for the carrier version of the F-111, a program the Navy abandoned. The Air Force, which continued on with the F-111 program, did not continue development of the AGM-53, and it was eventually shelved.

What we do know is that it was a TV guided version of the French Martel missile and that it was bogged down with powerplant problems during development. It was to have a range of about 40 miles, and during tests it was reported to hit a target more than 55 miles away.

AGM-62 Walleye
Manufacturer: Martin Marietta
Native Users: USAF, USN, USMC
Foreign Users:
Variants: Walleye I, Walleye II
Launching Aircraft: A-4, A-7, F-4
Operating Period: 1966 to late 1990s
Mission: guided bomb
he Story: Development began in the early 1960s at the China Lake Naval Warfare Center as an inexpensive, TV-guided weapon system that could be attached to a standard issue Mk 83 1,000 lbs bomb. The Walleye development was successful and the weapon entered operational use in 1966.

A few years later, the NWC specialists further developed the guidance and control systems, and the result was the Walleye II, a TV-guided 2,000 lbs bomb.

Further development in the 1980s resulted in increased range and a datalink capability that allowed a second aircraft to guide the bomb after it was launched by the first. These extended range Walleye's are called the Walleye 1ER/DL and 2ER/DL, depending on the size of the bomb. These Extended Range/Data Link versions were still in the US Navy inventory until very recently .

The Walleye II ER/DL consists of the same basic hardware as the Walleye II, except for a larger wing. These larger wings enables the weapon to be launched from longer range to a target. The addition of the data link pod at the rear of the bomb and a joystick controller in the aircraft, allows for the weapon to be remotely steered to a specific target with pinpoint accuracy.

Walleye I undergoes wind tunnel testing at Arnold Engineering Development Center, TN. (USAF Photo)

An F-4 Phantom delivers a Walleye I. (USAF Photo)

A Walleye I mounted on an F-105 Thunderchief. (USAF Photo)

The Walleye I ER/DL is about 10.5 feet long while the Walleye II ER/DL is about 12.5 feet long. The tailspan of the Walleye II is about six feet across!

The tactics employed for the ER/DL versions are similar to those used by the AGM-135. Two aircraft are in a trail formation, with the launching aircraft several miles in front. This aircraft launches the weapon and turns and immediately leaves the area, taking evasive actions as needed. The datalink on the bomb is picked up by the trailing aircraft and he guides it to the target while remaining a good distance away.

AGM-64 Hornet
Manufacturer: North American Aviation
Native Users: USAF
Foreign Users: none
Variants: unknown
Launching Aircraft: F-100 and others of that era
Operating Period: 1964
Mission: anti tank and anti vehicle missile
The Story: One of the first attempts to develop an anti-armor weapon, the Hornet was developed in the early 1960s. Several contracts for development and testing of electro-optical guidance systems and captive carry tests were awarded in the mid 1960s. Again, not much information is available.

AGM-65 Maverick
Manufacturer: Hughes Missile System (prime contractor) and Raytheon Company (alternate source)
Native Users: USAF, USN, USMC
Foreign Users: Bahrain, Belgium, Denmark, Egypt, Germany, Greece, Iran, Israel, Italy, Jordan, Kenya, South Korea, Kuwait, Malaysia, Mexico, Morocco, Netherlands, New Zealand, Norway, Pakistan, Philippines, Portugal, Saudi Arabia, Singapore, Spain, Sudan, Sweden, Switzerland, Taiwan, Thailand, Tunisia, Turkey, Venezuela, Vietnam, Yugoslavia.
Variants: AGM-65A/B/D/E/F/G/H
Launching Aircraft: F-4, F-5, F-15E, F-16, F/A-18, F-111, A-4, A-6, A-7, A-10, P-3, AV-8B, A-10, A-37, AH-64, Viggen, Hunter
Operating Period: Entered U.S. service in 1972; still in service
Mission: Short and medium range, TV, Infrared or Laser guided missile. Initially developed as an anti-tank/hard target weapon
The Story: A wonderful weapon (as long as you are not sitting in a tank, that is) that is still very effective. The AGM-65 Maverick is a tactical, air-to-surface guided missile designed for close air support, interdiction, and defense suppression missions. It provides stand-off capability and high probability of strike against a wide range of tactical targets, including armor, air defenses, ships, transportation equipment, and fuel storage facilities.

The Maverick is a modular design weapon. A different combination of the guidance package and warhead can be attached to the rocket motor section to produce a different weapon. The Maverick has three different seekers and two different warheads. The solid-rocket motor propulsion section is common to all variants. The

An F-15 Eagle fires an AGM-65 Maverick during testing. (USAF Photo)

Maverick 1 mounted to an underwing. (USAF Photo)

A U.S. Army Cobra helicopter fires an air-to-ground Maverick missile. (U.S. Navy Photo)

seeker options are electro-optical (EO) imaging, imaging infrared (IR), or a laser guidance package. The warhead is in the missile's center section. Either a 125-pound shaped-charge warhead or a 300-pound penetrator warhead can be used. A contact fuse in the nose fires the shaped-charge warhead. The penetrator uses a delayed-fuse, allowing the warhead to penetrate the target with its kinetic energy before firing. The latter is very effective against large, hard targets.

The original Maverick was a TV guided missile used for hunting tanks and armored vehicles. Once the pilot selected the target using the Maverick's TV, it was launched and the pilot could monitor the weapon's progress via the TV datalink. This original version was fairly short range, only a couple of miles, but as the weapon developed over the years it acquired new forms of guidance packages (imaging infrared and laser), larger warheads, and greater range.

Maverick B models have an electro-optical television guidance system. After the protective dome cover is automatically removed from the nose of the missile and its video circuitry activated, the scene viewed by the guidance system appears on a cockpit television screen. The pilot selects the target, centers cross hairs on it, locks on, then launches the missile. The Maverick B also has a screen magnification capability that enables the pilot to identify and lock on smaller and more distant targets.

The Maverick D has an imaging infrared guidance system, operated much like that of the A and B models, except that infrared video overcomes the daylight-only, adverse weather limitations of the other system. The infrared Maverick D can track heat generated by a target and provide the pilot a pictorial display of the target during darkness and hazy or inclement weather.

The Maverick E model is the only version having the laser-guided seeker section. It uses the heavyweight penetrator warhead. The U.S. Marine Corps are the only users of this variant.

The Maverick F is a naval variant of the D/G model (IR) currently in use by the U.S. Navy. It also uses the 300-pound penetrator warhead.

The Maverick G model essentially has the same guidance system as the D, with some software modifications that track larger targets. The G model's major difference is its heavyweight penetrator

The business end of the AGM-65 Maverick. (USAF Photo)

Right: An F-15E fires a Maverick at Nellis AFB, Nevada. (USAF Photo)

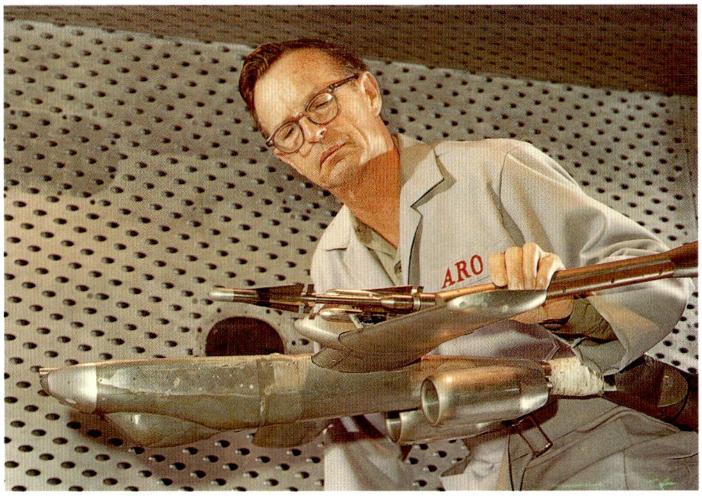

AEDC dual-purpose wind tunnel testing in 1975 matched the A-10 close support aircraft with the television-guided Maverick air-to-ground missile. (AEDC Photo)

AGM-69 Short Range Attack Missile being loaded onto a B-1 for fit and function testing. (USAF Photo)

warhead, while Maverick B and D models employ the shaped-charge warhead.

Maverick K models are currently in development. They were developed by taking a G model and replacing the IR guidance system with an electro-optical (EO) television guidance system. The final version, the -65H, carried a 250 lb. warhead, used active radar to lock onto its target, and had a range of up to 20 miles.

The infamous "Road of Death" that effectively ended the Gulf War was a result of a lot of Mavericks and 30-mm A-10 rounds.

The Maverick remains in service in a number of Air Forces because of its great reliability, ease of use, and effectiveness in killing armored targets.

AGM-69 Short Range Attack Missile (SRAM)

Manufacturer: Boeing
Native Users: USAF
Foreign Users: None
Variants: AGM-69A/B
Launching Aircraft: B-52, FB-111
Operating Period: 1972-1990
Mission: Long range, nuclear attack
The Story: The SRAM was an important part of the U.S. nuclear deterrence policy during the height of the Cold War. About 1,500 of these missiles, each topped with a 170 kiloton nuclear warhead, stood alert at numerous Strategic Air Command bases in the United States.

Close-up of a B-52 inboard pylon installation carrying six SRAMs. (USAF Photo)

Artist concept of the SRAM-T air-to-surface training mission. (McDonnell Douglas Photo)

A big missile, the SRAM topped the scales at over 2,000 lbs. As big as it was, the B-52 could still carry 20 of these deadly weapons, and the FB-111 carried six.

A range of only 150 miles or so meant that these bombers had to get in close to their targets if they were to be used. It could be set to detonate on impact, or as an air burst over a selected target.

The AGM-78 Standard Anti-Radiation Missile. (U.S. Navy Photo)

AGM-78 Standard Anti-Radar Missile

Manufacturer: Hughes Missile Systems
Native Users: USN, USAF
Foreign Users: Israel, South Korea
Variants: AGM-78A/B/C/D
Launching Aircraft: F-4, A-6, F-105
Operating Period: 1969-
Mission: Medium range, anti-radar
The Story: A precursor to the AGM-88 HARM, the Standard ARM attacked and destroyed ground-based radar sites. The Standard ARM checked in about 14 feet long, and weighed over 1,200 lbs. with a 30 mile range.

F/A-18 fires an AGM-78 Standard ARM. (U.S. Navy Photo)

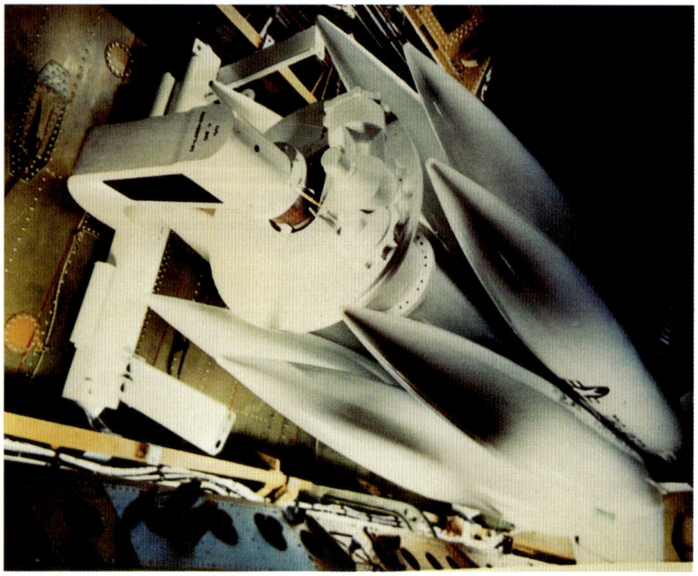

Left: The internally-carried rotary launcher in a B-52 could carry six SRAMs. (USAF Photo)

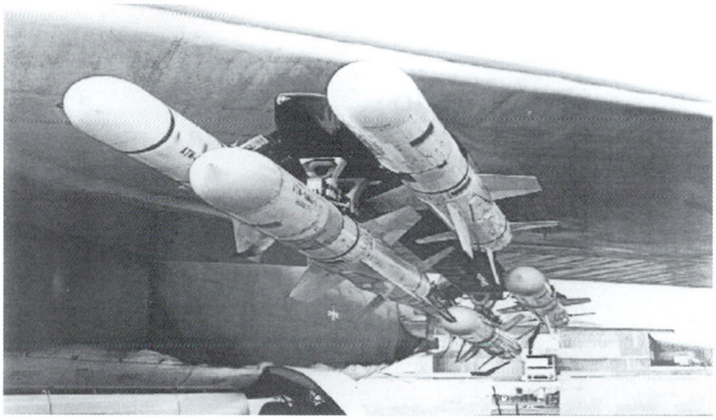

A trio of AGM Harpoon anti-shipping missiles on a B-52 in this undated photo. Most likely, the photo is from Loring AFB, Maine, one of the first B-52 units to take on the anti-shipping mission. (USAF Photo)

Two AGM-84 Harpoons on a Navy F/A-18. (U.S. Navy Photo)

AGM-84 Harpoon

Manufacturer: McDonnell Douglas

Native Users: USAF, USN

Foreign Users: Australia, Canada, Iran, Israel, Japan, South Korea, Kuwait, New Zealand, Portugal, Thailand, Turkey, and UK

Variants: AGM-84A/D/-84E, SLAM/-84H, SLAM-ER, GRAND SLAM

Launching Aircraft: A-6, A-7, B-1, B-52, F-16, F/A-18, F-20, Nimrod, P-3, S-3. For SLAM launches, A-6, B-52, F/A-18, F-27, F-111

Operating Period: 1977-current

Mission: Long range, anti-ship missile, long range, land attack missile (SLAM)

The Story: Originally designed as an anti-ship missile to be launched by B-52s and FB-111s, the Harpoon grew into a family with the development of ship- and submarine-launched versions. This original version was quite capable, and later grew even more capable with the addition of another fuel tank and the ability to be launched to an area and then search for a target.

A major mission change came in 1986, when McDonnell Douglas, the prime contractor, developed a land attack version, the SLAM (Standoff Land Attack Missile). The SLAM dumped the active radar seeker and incorporated the Imaging Infra-Red seeker from the Maverick. It also added a GPS guidance system and a datalink that allowed the transmission of the television image. The SLAM became operational in 1988.

The SLAM saw limited service in the Gulf War, primarily because it was still in the operational test and evaluation phase. How-

The AGM-84 in its sea-skimming mode. (U.S. Navy Photo)

Harpoon under the right wing of this F/A-18, along with AIM-9s and an AIM 120. (U.S. Navy Photo0

ever limited its action, it was successful enough to warrant an immediate order for an upgraded version, the SLAM-ER (Standoff Land Attack Missile-Enhanced Response). Now with an improved warhead and numerous software improvements, the SLAM-ER has a range of about 75 miles.

An even more advanced version, the GRAND SLAM, has even greater range (up to 200 miles) and an 850 pound penetrator warhead.

The sea and air-launched versions of the AGM-86 all carry a 500 lb high explosive, fragmentation warhead.

AGM-86 Air Launched Cruise Missile (ALCM)
Manufacturer: Boeing
Native Users: USAF
Foreign Users: None
Variants: AGM-86A/B/C
Launching Aircraft: B-52G/H, B-1, B-2 (????)
Operating Period: 1984-current
Mission: Long range, air launched cruise missile
The Story:
The ALCM's roots can be traced back to the SCAD decoy program of the 1970s, but it really took off in the late '70s when the USAF wanted a nuclear-capable cruise missile with a 1,500 mile range.

In February 1974 the Air Force entered into contract to develop and flight-test the prototype AGM-86A air-launched cruise missile, which was slightly smaller than the later B and C models. The 86A model did not go into production. Instead, in January 1977, the Air Force began full-scale development of the AGM-86B, which greatly enhanced the B-52's capabilities

The AGM-86B air-launched cruise missiles and AGM-86C conventional air-launched cruise missiles were developed to increase the effectiveness of B-52H bombers.

The ALCM was to be the weapon of choice for the B-52 fleet, to be used as a stand off, nuclear attack missile against the Soviet Union Russia. They could blast open a path for radar evading B-1s carrying other nuclear weapons.

A Conventional Air Launched Cruise Missile is loaded on a B-52 at RAF Fairford, UK, during Operation Allied Force. The CALCM carries a three thousand pound warhead. (USAF Photo)

The small, winged AGM-86B/C missile is powered by a turbofan jet engine that propels it at sustained subsonic speeds. After launch, the missile's folded wings, tail surfaces, and engine inlet deploy. The AGM-86B is then able to fly complicated routes to a target through use of a terrain contour-matching guidance system. The AGM-86C uses an onboard Global Positioning System (GPS) coupled with its inertial navigation system (INS) to fly. This allows the missile to guide itself to the target with pinpoint accuracy.

AGM-86B/C missiles increase flexibility in target selection. AGM-86B missiles can be air-launched in large numbers by the bomber force. B-52H bombers carry six AGM-86B or AGM-86C missiles on each of two externally mounted pylons, and eight internally on a rotary launcher, giving the B-52H a maximum capacity of 20 missiles per aircraft.

Later, in 1986, the USAF rebuilt a number of ALCM with conventional warheads, cleverly called the Conventional Air Launched

An AGM-86C Conventional Air Launched Cruise Missile in flight. The CALCM can be used for precision strikes against hard and deeply buried targets well behind enemy lines. (Boeing Photo)

A CALCM, a modified AGM-86, dives toward target impact during testing with the USAF. The CALCM achieved accuracy to within nine feet of target during these tests. (Boeing Photo)

Six AGM-86B ALCMs are carried on a pylon launcher on this B-52H. (Boeing Photo)

An airman secrues an AGM-86B Conventional Air Launched Cruise Missile (CALCM) aboard a B-52H Stratofortress at RAF Fairfield, UK. (USAF Photo)

Cruise Missile, or AGM-86C CALCM. It carries a 3,000 lb high explosive warhead up to 600 miles. In addition to the conventional warhead, the -86C also replaced the B model's terrain contour-matching guidance system and integrated a GPS capability with the existing inertial navigation computer system.

A -86D program is planned. This variant will retain the conventional warhead of the -86C and significantly improve its accuracy.

The CALCM became operational in January 1991 at the onset of Operation Desert Storm. Seven B-52s, from Barksdale AFB, LA, launched 35 missiles at targets in Iraq.

Boiling it all down, the ALCM, no matter which version, is a small, self-guided, unmanned bomber. Its radar reads the terrain, compares it to maps stored in its electronic brain, and makes corrections as needed on the way to its preprogrammed target. During its up to 1,500 mile-long flight, the missile flies between 25' and 500' to avoid radar.

Boeing began production in 1980, and six years later, when the production line closed, they had built 1,760 ALCMs.

AGM-88 HARM

Manufacturer: Texas Instruments/Raytheon
Native Users: USAF, USN, USMC
Foreign Users: Australia, Finland, Germany, Israel, Italy, Pakistan, Spain, and Turkey
Variants: AGM-88A/B/C
Launching Aircraft: A-6, A-7, EA-6B, F-4G, F-16, F/A-18, Tornado, F-111C
Operating Period: 1984-current
Mission: medium range anti-radar missile
The Story: The HARM is a second generation anti-radar missile, replacing the SHRIKE and Standard ARM as the weapon of choice. It has a higher speed, and greater accuracy and range than its predecessors.

A USAF F-16 fires an AGM-88 High Speed Anti-Radiation Missile during Red Flag, an annual Air Force exercise at Nellis AFB, NV. (USAF Photo)

A U.S. Marine Corps EA-6B Prowler, with an AGM-88 HARM and a jamming pod under its wing, readies for a mission over Kosovo during Operation Allied Force. (USAF Photo)

The HARM is highly effective in soft kills of SAM sites, killing the radar, but not the missile or its launcher.

With the retirement of the F-4G Wild Weasel from the USAF, the HARM shooting role has fallen to the EA-6B Prowler and the F-16CJ. The HARM was extensively used in the Gulf War, in the air war over Kosovo, and in Operations NORTHERN and SOUTHERN WATCH.

AGM-114 Hellfire
Manufacturer: Rockwell International
Native Users: USA, USMC
Foreign Users: Canada, Egypt, Greece, Israel, South Korea, Saudi Arabia, Sweden, and United Arab Emirates
Variants: AGM-114A/B/C/F/K, Hellfire 2, Longbow
Launching Aircraft: Attack helicopter, including AH-1 Cobra/Super Cobra, AH-64 Apache, and RAH-66 Comanche
Operating Period: 1985-current
Mission: Short range, laser guided air-to-surface attack
The Story: The Hellfire is the main air-to-ground missile of many of today's U.S. Army helicopters, including the AH-64 Apache gunship. The name comes from "Helicopter-Launched FIRE-and-forget missile." The Hellfire is primarily used in an anti-tank role.

Hellfire was born of the AGM-64 Hornet, a technology demonstrator built between 1963-66 by North American Rockwell. It proved the concept of using an optically-guided "fire and forget" air-to-surface missile to destroy heavily armored targets. From this successful concept, the Hellfire was first test-fired in 1971. But it was years before it made it into the Army inventory. The first Army unit to reach operational status was not until 1985!

After the Hellfire is launched, a laser designator must be brought to bear on the target by either the firing aircraft or a ground-based unit. When the laser seeker in the nose of the Hellfire finds the laser reflections, it simply follows the laser light to the target. Just before the Hellfire reaches its target, the missile pitches up sharply, then dives onto the target, a tactic which preys upon the comparatively thin armor on the tops of tanks.

The Hellfire opened offensive action in Desert Storm, as Apache gunships (each with a standard load of 16 Hellfires) took out Iraqi early-warning radar sites to clear a path for the air assault. Over 500 Iraqi tanks met their fate at the hands of the Hellfire missile.

The Hellfire 2 version incorporates a dual-warhead to attack and defeat today's reactive armor. For guidance, it combines an imaging infrared seeker and millimeter radar seeker. It also has an increased range—up to 6 miles.

AGM-119 Penguin
Manufacturer: Norsk Frosvarsteknologi (Northrop is U.S. subcontractor)
Native Users: USAF, USN,
Foreign Users: Greece, Norway, Singapore, Turkey
Variants: AGM-119A/B
Launching Aircraft: F-16, S-70 helicopter
Operating Period: late 1980s-mid 1990s
Mission: Medium range, infrared guided, air-to-surface missile
The Story: Designed by a Norwegian company as a ship-to-ship missile, the Penguin has undergone numerous changes. In addition to the ship-to-ship version, air-to-ship and air-to-surface versions are also operational.

The AGM-119A is the Penguin 3, a medium range missile with a range of about 30 miles when launched from higher altitudes. It carries a warhead of about 300 lbs. The AGM-119B version equates to the Penguin 2, and has a range of about 15-18 miles.

AGM-122 Sidearm
Manufacturer: Motorola
Native Users: USN, USMC, USA
Foreign Users: none
Variants: AGM-122A/B
Launching Aircraft: AV-8B, A-4, AH-64, AH-1 SeaCobra, F-4G, F/A-18,
Operating Period: 1989-current
Mission: short range, anti-radar missile

AN AGM-119A Penguin Anti-Ship Missile is launched from an F-16 Fighting Falcon aircraft during testing. The F-16 is also armed with AIM-9 Sidewinder missiles. (USAF Photo)

The Story: The Sidearm is a short-range, anti-radar missile designed to attack SAM sites. It was developed as an inexpensive missile to be used by U.S. Marine Corps fixed-wing and helicopter aircraft. It is basically an AIM-9C "Sidewinder," the only passive radar-guided Sidewinder produced. The semi-active seeker of the AIM-9C was replaced by a wideband passive radar seeker to give the Sidearm capability against a wide range of radar installations.

Its biggest advantage is that any aircraft capable of carrying a Sidewinder can use the Sidearm.

AGM-123 Skipper II

Manufacturer: Electronics and Space Corporation
Native Users: USN
Foreign Users: none
Variants: AGM-123A
Launching Aircraft: A-6, A-7 (neither in service any longer), and F/A-18
Operating Period: 1985-current
Mission: Laser guided, short range missile
The Story: The AGM-123 "Skipper II" is a laser-guided standoff anti-ship missile based on existing missile and bomb components. The body of the Skipper is a standard Mk 83 1,000 lb general purpose bomb. The rocket motor is derived from the AGM-45 Shrike anti-radar missile. The seeker is based on the laser-guided Paveway II series of guided precision bombs. The first combat seen by the Skipper was on April 18, 1988, when a U.S. Navy A-6 Intruder struck the Iranian frigate *Sahand* with an AGM-123 after the ship engaged U.S. Navy ships in the Strait of Hormuz. The *Sahand* was also hit by an AGM-84 Harpoon and a surface-launched RGM-84. The *Sahand* did not survive this engagement. The Skipper was initially operational in 1985.

AGM-129 Advanced Cruise Missile

Manufacturer: General Dynamics
Native Users: USAF
Foreign Users: none
Variants: AGM-129A/B
Launching Aircraft: B-1, B-2, B-52
Operating Period: 1991-current
Mission: Long range cruise missile
The Story: The original stealth cruise missile, the AGM-129, was originally seen as the replacement for the AGM-86 ALCM. The -129 was designed with longer range, better accuracy, and a lower radar cross section.

The missile is wedge shaped, with forward swept winglets that deploy after being launched. The -A model carries a nuclear warhead, while the -B version is conventional.

McDonnell Douglas (now Boeing) was contracted as a second source in 1987. The Air Force originally planned on buying about 1,000 of these missiles from the two contractors, but later cut that

AGM-129, the Advanced Criuse Missile, in flight. Notice the unusual configuration with the vertical stabilizer on the bottom and the forward-swept wings. (USAF Photo)

A side view of the AGM-130. (USAF Photo)

number to 460. Production ended in 1993. A victim of the peace dividend? While the current U.S. leadership's preference for firing off cruise missiles at suspected terrorists (Sudan, Afghanistan, and Serbia), the inventory of cruise missiles continues to drop. The Air Force and the Navy are both retrofitting some nuclear carrying missiles into conventional missiles.

AGM-130
Manufacturer: Boeing (original contractor Rockwell International)
Native Users: USAF
Foreign Users: South Korea
Variants: AGM-130A/B/C
Launching Aircraft: F-15E, F-111F
Operating Period: 1992-current
Mission: Guided standoff attack
The Story: Take an Mk84 2,000 lb bomb (for -130As), submunition dispenser (for -130Bs), or BLU-109 penetrating warhead (-130Cs) and strap on a guidance kit and a solid propellant rocket motor, and you have got a very deadly stand off weapon with a range of up to 25 miles. One of the largest weapons in sheer size, the 3,000 lb, nearly 13' long AGM-130 is being used very successfully in theaters around the world.

The AGM-130 began development in 1984 and went into operational service in 1993. The AGM-130 employs a rocket motor for extended range and an altimeter for altitude control. It provides a significantly increased standoff range than the GBU-15, which it is based on.

The AGM-130 has two variants, based on the warhead. The AGM-130A is an MK-84 blast/fragmentation warhead, and the AGM-130C is a BLU-109 penetrator.

The AGM-130 is equipped with either a television or an imaging infrared seeker and data link. The seeker provides the launch aircraft a visual presentation of the target as seen from the weapon. During free flight this presentation is transmitted by the AXQ-14 data-link system to the aircraft cockpit monitor.

The seeker can be either locked onto the target before or after launch for automatic weapon guidance, or it can be manually steered by a weapon systems officer. Manual steering is performed through the two-way data link.

The AGM-130 is designed for use in the F-15E aircraft. Development of the AGM-130 began in 1984 as a product improvement of the GBU-15 guided glide bomb.

Basic tactics for launch include having the aircraft fly to a prebriefed launch position. The survivability of aircraft and crew is enhanced by launching the weapon at low altitude and significant standoff range, thus avoiding detection by enemy air defenses. After launch, the weapon flies through glide-powered-glide phases toward the target area with midcourse guidance updates provided by global positioning system (GPS) navigational information, or by the weapon systems officer through the data link.

Upon termination of the powered flight phase the rocket motor is ejected. As the target comes into view, the weapon systems officer has dual flexibility in guiding the weapon via the data link. For automatic terminal homing, the guidance tracker is locked on target but can be manually updated for precision bombing. When total manual guidance is used, the operator manually guides the weapon to the target aimpoint. For those aircraft not equipped with a data-link pod, the weapon may be launched in the direct attack mode.

The AGM-130 was used very successfully in the air war over Serbia. Using the TV guided version, because of its great accuracy, it was used against a number of bridges.

AGM-131 Short Range Attack Missile II
Manufacturer: Boeing
Native Users: USAF
Foreign Users: possible UK sales/coproduction
Variants: AGM-131A/B SRAM II/SRAM-T
Launching Aircraft: B-1, B-2, F-16, F-15E, FB-111, Tornado
Operating Period: Canceled before reaching operational status
Mission: Long range, nuclear air-to-surface missile
The Story: SRAM II (Short Range Attack Missile II) had a seven year development life, from 1985 to 1992. The SRAM-T had but one year of life, 1990-1991. It was to replace the AGM-69, a missile with fewer parts, higher reliability, and lower maintenance requirements.

A head-on view of the AGM-130. (USAF Photo)

Artist's concept of the AGM-132 SRAM II air-to-ground nuclear missile. (USAF Photo)

Its design also allowed for easy swap out of warheads, allowing for the missile to carry either a nuclear or conventional warhead.

The program was canceled in 1991, with no follow-on program identified.

AGM-136 Tacit Rainbow
Manufacturer: Northrop, Raytheon (second source)
Native Users: USAF, USN
Foreign Users: none
Variants: none
Launching Aircraft: B-52, F-111, A-6
Operating Period: none, program terminated in 1990
Mission: medium range, anti-radar missile
The Story: Development began in 1984 on this low-cost, anti-radar missile with the ability to be launched and loiter in the area of a target, then lock on and attack a radar once it begins radiating.

The idea was to relieve some of the pressure on the Wild Weasel community by being able to launch a number of these missiles ahead of the attacking force. Once the radars came up to launch surface-to-air missiles, the Tacit Rainbows would attack without having to put pilots and aircraft in danger, as would happen if the Wild Weasels led the strike package.

AGM-137A TSSAM
Manufacturer: Northrop
Native Users: None, canceled
Foreign Users: None
Variants: none
Launching Aircraft: None.
Operating Period: None
Mission: Stand off, precision land attack
The Story: Tri Service Stand Off Attack Missile, funded as a black program in 1986, it was canceled in 1994 after $3 billion was spent on the development of a stealthy missile that could be used by all services. The program reportedly had problems in flight testing.

Tacit Rainbow in a B-523H rotary launcher. (USAF Photo0

An artist's rendering of the AGM-142 in flight. (USAF Photo)

The AGM-142, or more commonly the Haqve Nap, is an Israeli design. Here, it is shown on the wing of a B-52. (USAF Photo)

AGM-142 Raptor
Manufacturer: Lockheed Martin and Rafael
Native Users: USAF
Foreign Users: Australia, South Korea
Variants: AGM-142A/B/C/D
Launching Aircraft: B-52
Operating Period: 1992-current
Mission: Medium range, stand off attack
The Story: Formerly called Have Nap, the AGM-142 is an American version of the Israeli "Popeye" weapon system. The AGM-142 is manufactured by a joint venture company being established between Lockheed Martin and Rafael, an Israeli firm.

AGM-142 is a standoff, air-to-ground, electro-optical guided missile fitted with either a television (TV) or imaging infrared (IIR) seeker. The system also includes an INS and data link pod (DLP).

The missile weighs 3,000 lbs, is powered by a single stage solid rocket motor, and armed with either a general purpose 750 lb fragmentation warhead or an 800 lb penetrating warhead.

It began development in 1988 and went operational in 1990. The B-52 can carry four Raptors, each with a 750 lb warhead. The missile uses IR or TV guidance systems and has a range of about 55 miles.

Artist's drawing of the AGM-154 Joint Stand-Off Weapon. (DOD Photo)

AGM-154 (Joint Stand Off Weapon)
Manufacturer: Raytheon
Native Users: USAF, USN, USMC
Foreign Users: UK
Variants: AGM-154A/B/C
Launching Aircraft: F-15, F-16, F/A-18, F-117, AV-8B, B-1, B-2, Tornado, Jaguar
Operating Period: 1990s-current
Mission: Precision land attack
The Story: A winged, guided but unpowered bomb, it first saw operational use in Iraq in February 1999. It has an integrated GPS/INS seeker. The key advantage to the JSOW is its flexibility. It can be configured to carry submunitions, or a single high explosive penetrating warhead. Though it is designed to be unpowered, a strap on rocket motor is reportedly being developed to allow for even longer stand off distances. Even in its unpowered state, it has great glide characteristics that allow it to hit targets from as far away as 40 miles. With a rocket motor strapped on, it should be able to take out targets more than 100 miles away, with deadly precision.

The AGM-154 is intended to replace the Maverick, Walleye, and Skipper missile systems. The -154 will be able to attack targets day or night, and in adverse weather.

It began development in 1989 and reached operational status in 1998. The U.S. military plans on buying nearly 25,000 of these weapons.

The JSOW weighs about 1,500 lbs, with the warhead accounting for about 1,000 lbs of that weight.

The B-2 Spirit can carry 16 JSOWs, hitting 16 different targets on a single pass. Other aircraft can carry varying numbers of JSOWs, depending on the aircraft weapons/fuel tank configuration.

Have Slick
Manufacturer:
Native Users: USAF
Foreign Users: None
Variants: None
Launching Aircraft: F-117, B-2, F-22
Operating Period: None
Mission: Family of submunition carriers
The Story: A black program for many years, details first came out in the early 1990s. Have Slick may be a family of submunition carriers with a range of about 25 miles. The main body, which carries the submunitions, is about 14 feet long, and roughly shaped like a rounded over triangle. It is believed to weigh in the neighborhood of about 3,000 lbs, with about 2,000 lbs devoted to the explosive parts of the weapon.

The Joint Air-To-Surface Stand-Off Missile on an F-16 during testing. The F-16, from Eglin AFB, FL, has AIM-9 Sidewinders on the wingtip pylons. (USAF Photo)

JASSM

Manufacturer: Lockheed Martin (Boeing as second source)
Native Users: USAF, USN
Foreign Users: None
Variants: None
Launching Aircraft: B-52, F-16, F/A18 initially, with follow on testing for B-1 and B-2
Operating Period: Will reach initial operational capability in 2003
Mission: Long range, conventional air to surface cruise missile.
The Story: Still undergoing testing, the Joint Air to Surface Standoff Missile will be able to attack a variety of both fixed and relocatable targets using conventional munitions. The first launch of the JASSM was in August 1999 at White Sands Missile Range, NM. It was an unpowered test to evaluate aircraft separation, deployment of the wings and tail surfaces, and basic flight maneuvers. The test was successful.

The JASSM is the next generation long range cruise missile that uses advanced global positions system/inertial navigation system guidance. The USAF/USN have committed to buying nearly 2,500 JASSMs, and have identified a need for another 1,500 that are yet to be funded.

The JASSM weighs in at over 2,200 lbs, with 1,000 lbs of that being the warhead. It will have a range of at least 175 miles.

One of the biggest differences between JASSM and the AGM-86C that it will replace is cost. The -86C CALCM costs approximately $1.5 million per copy, while the JASSM comes in at under $400,000 per missile.

The JASSM in flight test, launched from an F-16D at Edwards AFB. (USAF Photo)

Gravity Weapons

With the advent of the GPS system, laser designators, and the like, what used to be called "dumb bombs" are now among the most accurate in history. In the '60s, a circular error of probability (the diameter in which you could expect a bomb to fall) was on the order of hundreds of feet. Now manufacturers are talking about feet. The unpowered JDAM used so successfully in the air war with Serbia uses a GPS guidance system that can put a 2,000 lb bomb into a shopping basket from thousands of feet up and miles away.

The basic building block of the unpowered, precision guided munitions is the Mk. 80 series of general purpose bombs. Developed first in the 1950s, the GP bombs are a family of low-drag, general purpose bombs.

With the advent of the jet fighter, the need for a bomb that minimized drag was also needed. Early fighters, such as the F-86, F-100, F-4, and even today's F-16 and F-15 all carry bombs externally. The result is the Mk. 80 series of bombs that can be seen all around the world.

The Mk. 80 family has bombs ranging in weight from 250 lbs to 3,500 lbs. Despite the vast difference in weight, they are relatively small in design, with the only changes being in the scale of the bomb. The Mk. 80 series of bombs are the basic building blocks of the Paveway family of bombs now in use around the world by a number of U.S. allies.

The Paveway was the result of a long known fact that the free-fall, or "iron" bomb is a cheap weapon that is also highly effective, but only if it hits its target. Early experience in the Vietnam War persuaded the U.S. Air Force that even its most advanced sights were inadequate to provide the free-fall bomb with the terminal accuracy for the destruction of planned targets without a huge expenditure of bombs and large numbers of sorties. It would be the daylight bombing raids of World War II all over again, with the loss of large numbers of highly trained aircrew and their valuable warplanes. The Air Force did not want to go there again.

A B-1 bomber is loaded with 500 pound bombs in preparation for another mission over Kosovo during Operation Allied Force. (USAF Photo)

A Navy F/A-18 carries a laser guided bomb on its left wing during an exercise at NAS Fallon, NV. (McDonnell Douglas Photo)

An airman drives the bomb cart, carrying a 500 pound bomb, to a waiting B-1 Lancer bomber at RAF Rairford, UK, during operation Allied Force. (USAF Photo)

During the mid-1960s considerable effort was spent in creating a program to improve the accuracy of free-fall bombs. One of the key elements in this program had been started in 1965 by the Armament Development and Test Center at Eglin AFB, FL, with the partnership of the Mk. 80 series manufacturer, Texas Instruments. The plan was to create a cheap system for giving any of the Mk 80 series (and other tactically important free-fall bombs) weapons a higher degree of accuracy by adding a semi-active laser homing system and a control package.

This program led to the "Paveway" series of laser-guided bombs, of which the first were introduced in 1967 after the initial release of a laser-guided development weapon in 1965.

The kit could be fitted to standard low-drag bombs for maximum accuracy when used with an air- or ground-based laser designator. The "Paveway I" soon proved in operations from 1968 that its primary benefit, other than the very high degree of terminal accuracy, was that the launch warplane needed no modification to carry the weapon (any warplane capable of lifting the baseline "iron" bomb could deliver a "Paveway I").

In all cases the guidance unit was the same, the differences between the various systems lying in their attachments and variously enlarged fins. The guidance system drew current from a thermal battery activated at launch, and was steered by canard control fins powered by a hot gas generator.

In 1980 an improved series of Paveway bombs was developed, based on the same basic concept as the earlier ones, but with a simpler (and cheaper) guidance.

The weapons within this important basic designation are:

- GBU-10C/D/E/F "Paveway II" based on the 2,000 lb Mk 84 bomb
- GBU-10G/H/I/J "Paveway II" based on the 2,000 lb BLU-109/B Improved 2,000 lb Penetrating Warhead
- GBU-12B/C/D "Paveway II" based on the 500 lb. Mk 82 bomb
- GBU-16A/B "Paveway II" based on the 1,000 lb Mk 83 bomb
- Mk 13/18UK "Paveway II" based on the 1,000 lb British Mk 13/18 bomb

In each case the weapon functions in a manner basically similar to that of the "Paveway I" series.

In 1987, an improved version of the "Paveway II" intended mainly for use in the lousy weather and high-threat scenario of European operations was created. The new series of bombs were fitted with high-lift folding wings at the rear, a microprocessor-based control system that resulted in increased range and accuracy, and a digital and an improved scanning seeker.

The "Paveway III" is also designated the Low-Level Laser-Guided Bomb or Laser-Guided Penetrator Weapon, and was designed for release at low level (in either level flight or a zoom climb),

A U.S. Navy F/A-18 with a full load of Mk-83 gravity bombs. (U.S. Navy Photo)

A USAF F-15E drops a full load of Mk-83 gravity bombs. (McDonnell Douglas Photo)

and are fitted with high-lift folding wings. Paveway III bombs can also be released at higher altitudes in dives as steep as 60°. The result is a weapon of far greater operational flexibility than the previous Paveway series of bombs.

The "Paveway III" series of bombs allow for low-level and off-axis stand-off delivery in low visibility, with the advantages of a considerably expanded launch envelope.

When looking at the gross weight of a bomb, about half the weight of the bomb is the actual explosive, the rest being the casing, mounting hardware, fins, and fuzes.

The most important element of all is the fuze. Today's target planners not only need to decide what target to hit, but they need to know which type of weapon and fuze to use to obtain the results they are looking for. Fuzes today can vary from one that detonates immediately on impact to one that can be programmed to delay for a specific period of time. Today's research is even looking at fuzes that can count the number of floors it penetrates in order to detonate not just on the building targeted, but on the exact floor you want to destroy! Other fuzes are proximity fuzes that detonate at a specified altitude, or ones that sit and wait for a large mass of metal (like a tank) to come within lethal range.

The penetrating bomb came into its own during Desert Storm, when the U.S.-led coalition needed to put at risk Saddam Hussien's buried commanders. In a rapid development program that lasted mere weeks, the U.S. designed, tested, and shipped to the forces in Saudi Arabia the GBU-28, the "Bunker Buster." Specially designed with super hardened steel, the Bunker Buster was like a nail going through balsa wood. With its specially designed nose that could penetrate reinforced concrete, matched with a precision guidance kit, it was an incredibly accurate and deadly weapon.

Another development that began in Vietnam and was perfected 20 years later is the cluster bomb. Now the focus of a number of United Nations objections, the cluster bomb is incredibly effective against area targets or massed troops.

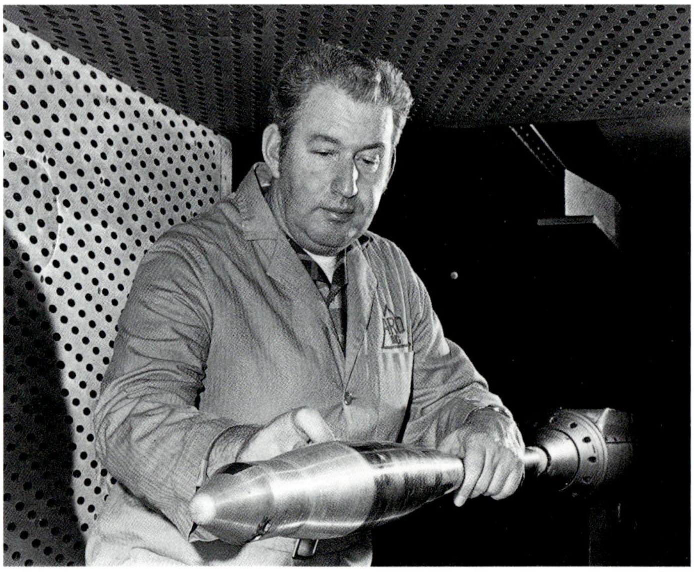

A scale model of an Mk-82 modified guided bomb is adjusted before wind tunnel testing. (USAF Photo)

Airmen at Aviaqno AB, Italy, prepare to load a GBU-12 Laser Guided Bomb on an F-15E during NATO's Operation Allied Force. (USAF Photo)

The Navy led the development of the Rockeye II. Adopted by both the Navy and the Air Force, it was particularly effective during attacks of SAM sites in Vietnam. The Rockeye II would drop nearly 250 submunitions, each with its own fuze, over an area whose size was determined by the altitude of the drop.

The Rockeye II was the most dropped weapon in Desert Storm, with nearly 28,000 dropped, according to the Department of Defense.

Today's U.S. military has three basic cluster bombs in its inventory. All three weigh in around 800 lbs, and all have the same function—anti-personnel, though they differ in the area of fuzing and warhead. The weapons can either be fragmentation or incendiary, and can either detonate on impact or time delay. Targeters determine the fuze and warhead based on the missions goal.

The next evolution of the CBU was to make it a stand off weapon. Often the target is so heavily defended that overflying it to drop cluster bombs would be suicidal. The Tactical Munitions Dispenser was the result. Checking in at around 1,000 lbs, the Air Force has three variants of the TMD in its inventory. Again, the targets are massed forces, either personnel or armored vehicles, and fuzing allows for immediate detonation or timed delays.

GBU-10/12/16 Paveway II
Manufacturer: Texas Instruments
Native Users: USAF, USN, USMC
Foreign Users: some 34 countries, including Australia, Canada, Greece, Israel, Netherlands, Saudi Arabia, South Korea, Spain, Taiwan, Turkey, UK, and USA.
Variants: GBU-10C/D/E/F/G/H/I/J Paveway II; GBU-12A
Launching Aircraft: too numerous to list
Operating Period: 1965-current
Mission: precision strike, laser-guided bomb
The Story: The GBU-10C/D/E/F "Paveway II" is based on the 2,000 lb Mk 84 bomb, while the GBU-10G/H/I/J "Paveway II" is based

on the BLU-109/B Improved 2,000 lbs Penetrating Warhead. The GBU-12 is based on the Mk. 82 500 lb bomb. The GBU-16 is based on the Mk-83 1,000 lb bomb

The beauty of these, and all the laser-guided GBU entries to follow, is that an aircraft carrying the bomb does not need to have any special systems installed to carry out precision strike missions. As long as one aircraft, or someone on the ground, can aim a laser at the target, the laser detector in the bomb's nose will "see" the laser (weather cooperating), and the pilot can drop it with a high degree of certainty of hitting the target. For these weapons, no special attachments are required, and there are no special electrical connections. Everything needed to hit the target is contained in the bomb.

As metioned, the Guided Bomb Unit-10 (GBU-10) uses a 2,000 lb general purpose or penetrating warhead. The operator illuminates a target with a laser designator, and then the munition guides to a spot of laser energy reflected from the target.

The munition was used during Operation Desert Storm and, according to the U.S. Air Force, hit 78 percent of its targets. In Operation Desert Storm, GBU-10s were used extensively by F-15Es and F-111Fs, mainly against bridges, Scuds, C3I (command, control, communications, intelligence) nodes, and bunkers. Of the 2,637 expended, over one third were dropped by F-111Fs, and the rest by F-117s, F-15Es, and Navy and Marine Corps aircraft.

The F-16 was the first single-seat fighter to achieve accurate, unassisted delivery of laser guided bombs. Here, an F-16 delivers a GBU-10 at 5,000 feet while traveling at more than 500 miles per hour. (USAF Photo)

A GBU-10. (Holder Collection)

There are two generations of GBU-10 LGBs: Paveway I with fixed wings and Paveway II with folding wings. Paveway II models also have improved detector optics and housing made of injection-molded plastic to reduce weight and cost, and an increased detector field of view. (Paveway II's instantaneous field of view is thirty percent greater than that of the Paveway I's field of view)

GBU-12

Manufacturer: Texas Instruments

Native Users: USAF, USN, USMC

Foreign Users: some 34 countries, including Australia, Canada, Greece, Israel, Netherlands, Saudi Arabia, South Korea, Spain, Taiwan, Turkey, UK, and USA.

Variants: GBU-10C/D/E/F/G/H/I/J Paveway II; GBU-12A

Launching Aircraft: A-6, A-7, A-10, B-52, F-14, F-15, F-16, F/A-18, F-111, F-117, and others too numerous to list

Operating Period: 1976-current

Mission: precision strike, laser-guided bomb

The Story: The GBU-12 is based on a 500-pound general purpose warhead. The operator illuminates a target with a laser designator, and then the munition guides to a spot of laser energy reflected from the target.

Computational Fluid Dynamics is emerging as a valuable tool in predicting airflow around weapons without having to conduct wind tunnel testing. Here, an F-15E with two GBU-12s, a LANTIRN pod, and an external fuel tank, as seen by CFD. (USAF Photo)

A rack of 500-pound bombs await loading into a nearby B-1 Lancer bomber. (USAF Photo)

Airmen at Aviano Air Base, Italy, secure the hooks to the eyelets on a GBU-12, before it was loaded into a flatbed cart. (USAC Photo)

The munition was used during Operation Desert Storm and hit 88 percent of its targets. During Desert Storm the GBU-12 was dropped by F-111Fs, F-15Es, and A-6s, mostly against fixed armor. It was the F-111F tank-busting weapon of choice. Of the 4,493 GBU-12s employed, over half were dropped by F-111Fs.

Like the GBU-10, there are two generations of GBU-12 LGBs: Paveway I with fixed wings and Paveway II, with differences as noted in the GBU-10 section.

GBU-15

Manufacturer: Boeing and Raytheon
Native Users: USAF
Foreign Users: none
Variants: GBU-15 V1/B, V2/B, -I
Launching Aircraft: F-111, F-15E
Operating Period: 1983-current
Mission: air-to-surface precision weapon
The Story: In essence a precision guided bomb, the GBU-15 uses either an Mk 84 (2,000 lbs) or a BLU-109 (2,000 lbs penetrator) warhead. The GBU-15 is an unpowered glide weapon used to destroy high value enemy targets. It was designed for use with F-15E, F-111F, and F-4 aircraft, but the U.S. Air Force is currently only deploying it from the F-15E. Its guidance system and wings give it a stand off range of many miles.

The two versions use the same Mk84 warhead, but differ in guidance systems. The V1 uses a TV, and the V2 uses infrared guidance.

The weapon can be employed either as a line of sight, where pilot/WSO locks a target into the guidance system before it leaves the jet, or he can drop it in an area and "fly" it to a target via datalink. The GBU-15 has the capability of using the buddy system, where one plane will drop the bomb and a buddy aircraft will pick up the datalink and fly it into the target from a safe standoff range.

The weapon consists of modular components that are attached to either an MK-84 general purpose or BLU-109 penetrating warhead. Each weapon has five components—a forward guidance section, warhead adapter section, control module, airfoil components, and a weapon data link.

A GBU-15 on an F-15E Strike Eagle. (USAF Photo)

A GBU-15. (Holder Collection)

The business end of the GBU-15 laser guided bomb. (USAF Photo)

The GBU-15 may be used in either a direct or an indirect attack. In a direct attack, the pilot selects a target before launch, locks the weapon guidance system onto it, and launches the weapon. The weapon automatically guides itself to the target, enabling the pilot to leave the area. In an indirect attack, the weapon is guided by remote control after launch. The pilot releases the weapon and, via remote control, searches for the target. Once the target is acquired, the weapon can be locked to the target or manually guided via the data-link system.

This highly maneuverable weapon has an optimal, low-to-medium altitude delivery capability with pinpoint accuracy. It also has a standoff capability. In Desert Storm, F-111F pilots used GBU-15 glide bombs to seal flaming oil pipeline manifolds sabotaged by Saddam Hussein's troops.

In summary, the GBU-15 is highly effective in destroying heavily defended targets due to its large warhead, great accuracy, and standoff capability.

GBU-16

Manufacturer: Texas Instruments
Native Users: USAF, USN, USMC
Foreign Users: Numerous countries, including Australia, Canada, Greece, Israel, Netherlands, Saudi Arabia, South Korea, Spain, Taiwan, Turkey, UK, and USA.
Variants: GBU-16
Launching Aircraft: A-6, A-10, F-14, F-15, F-16, F/A-18, F-111
Operating Period: 1976-current
Mission: precision strike, laser-guided bomb
The Story: The GBU-16 uses a 1,000 pound general purpose warhead. The operator illuminates a target with a laser designator, and then the munition guides to a spot of laser energy reflected from the target. The GBU-16 consists of an MK-83 1,000 pound bomb modified with a common Paveway II laser guidance kit. During Desert Storm virtually all 219 GBU-16s were dropped by Navy A-6s, which had the capability to laser the target themselves (self-designation).

FA/18 Hornet aircraft flying from USS *Enterprise* dropped GBU-16 laser guided bombs during the waves of attacks against Iraq in support of Operation Desert Fox in December 1998.

Paveway II ready for action. (USAF Photo)

Left: A GBU-15 glide bomb is prepared for wind tunnel testing. This particular bomb has a rocket motor attached, becoming an AGM-130. (USAF Photo)

Paveway III in flight. Note the significantly larger rear wings. This increase in size results in better maneuverability and greater range. (USAF Photo)

GBU-24 Paveway III
Manufacturer: Raytheon
Native Users: USAF, USN
Foreign Users: numerous
Variants: GBU-24A/B
Launching Aircraft: F-14, F-15E, F-16, and F/A-18, among others
Operating Period: 1986-current
Mission: precision air-to-ground penetrator bomb
The Story: The Paveway III combines the latest in laser guided technology with the advanced BLU-109 penetrating warhead to achieve a deadly and accurate bomb that can take out deeply buried, hardened targets.

It can be dropped from nearly any altitude, and its computer brain will fly the bomb on a course and angle of attack that will maximize the effectiveness of the penetrating warhead.

The GBU-24's Low Level Laser Guided Bomb [LLLGB] uses either a 2,000 pound MK-84 general purpose or BLU-109 penetrator bomb. The GBU-24 was developed in response to sophisticated enemy air defenses, poor visibility, and to counter limitations in low ceilings. The weapon is designed for low altitude delivery, and with a capability for improved standoff ranges to reduce exposure.

The GBU-24 has a low-level, standoff capability of more than 10 nautical miles. Its performance envelopes for all modes of delivery are improved because the larger wings increase maneuverability. Paveway III also has increased seeker sensitivity and a larger field of regard.

As with other GBUs, the operator illuminates a target with a laser designator, and then the munition guides to a spot of laser energy reflected from the target.

One way to deliver LGBs from low altitude is a loft attack. In this maneuver, the aircraft pulls up sharply at a predetermined point some miles from the target, and the LGB is lofted upward and toward the target.

It also has a larger delivery envelope for the dive, glide, and loft modes than do the earlier LGBs.

In the Gulf War all of the 1,181 GBU-24s were released by F-111Fs.

In 1996 the Navy conducted tests of the F-14 with the GBU-24B/B Hard Target Penetrator Laser-Guided Bomb at Naval Air Station Patuxent River, MD, as part of an air-to-ground development.

A U.S. Navy aviation ordnanceman applies the final wiring to a GBU-24 laser guided bomb attached to an F-14 aboard the *USS Theodore Roosevelt*. (U.S. Navy Photo)

U.S. Navy ordnancemen prepar a pair of GBU-24 laser guided bombs for an F-14 aboard the *USS Theodore Roosevelt*. (U.S. Navy Photo)

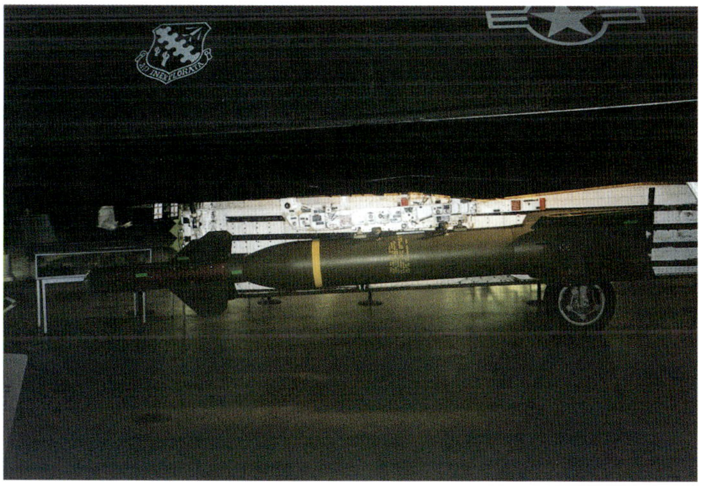

The GBU-27 used by the F-117 Nighthawk stealth fighter. (Holder Collection)

The GBU-28 "Bunker Buster" bomb. (Holder Collection)

GBU-27
Manufacturer: Lockheed Martin
Native Users: USAF
Foreign Users: none
Variants: GBU-27A/B
Launching Aircraft: F-117
Operating Period: 1987
Mission: precision strike, guided bomb
The Story: A modified GBU 24, the GBU-27A/B is used only by the F-117 Nighthawk. The modifications to the GBU-24 include a shortened warhead, slightly smaller wings, and adapting three guidance systems to the Nighthawk's targeting system. The shortening of the various parts of the GBU-24 was necessary to get the bomb to fit into the F-117's bomb bay. The changes also resulted in a bomb that has a shorter "throw" (range), but that is offset by the stealth jet's ability to fly closer to its target without being detected.

The GBU-27 was designed specifically for use by the F-117's advanced target acquisition/designator system. The GBU-27 uses a BLU-109 improved performance 2,000 lb bomb developed in 1985 under the project name HAVE VOID. The BLU-109 was designed for use against hardened structures and features a high-strength forged steel case and a new delayed-action tail fuze. It carries 550 pounds of high explosives and can penetrate more than six feet of reinforced concrete.

The GBU-27 uses a modified Paveway II guidance control for the ideal impact angle against various targets. For example, it will hit an aircraft shelter from the vertical, but makes a horizontal approach to a bridge support. A Paveway II tail assembly with folding wings completes the bomb.

The F-117 can carry two GBU-27s in its weapons bay, and is reportedly capable of hitting a one square meter target from an altitude of 25,000 feet.

Another view of the GBU-27. Note the jagged leading edge of the F-117 bomb bay door behind the bomb. (Holder Collection)

GBU-28B
Manufacturer: Lockheed Martin
Native Users: USAF
Foreign Users: none
Variants: none
Launching Aircraft: F-111, F-15E
Operating Period: 1991-current
Mission: precision strike, buried/hardened targets
The Story: The bunker buster of Desert Storm fame, the GBU-28 was developed, tested, and employed in a matter of weeks for use against Iraq's buried command and control bunkers. Initially, only four bombs were made, two for testing and two for use in Iraq. The testing went well, as did the bunker busting mission.

What makes the GBU-28 unique is the massive size of the bomb—nearly three tons! The GBU-28 is a 5,000 pound laser-guided conventional munition that uses a 4,400-pound penetrating warhead. The bombs are modified Army 8-inch artillery tubes, weigh 4,637 pounds, and contain 630 pounds of high explosives. They are fitted with GBU-27 LGB kits.

Called "Paveway III Deep Throat" for use by the F-15E Eagle and F-111F, the weapon was designed, produced, deployed, and used within a mere 17 days during Desert Storm in 1991. The GBU-28 was dropped from high altitude while the aircraft was in super-

A GBU-31/32 Joint Direct Attack Munition (JDAM) is dropped from a B-2 Spirit during tests.

sonic flight, and the resulting gravity-aided speed resulted in such kinetic energy at impact that "Deep Throat" could penetrate more than 20 ft of reinforced concrete or 100 ft of earth before detonation in command centers and bunkers

It was so effective that the Air Force contracted buy up to another 300, but with even more improved guidance systems and fuzing.

An artist's rendering of an F-22 delivering a GBU-32 (1,000 pounds) JDAM. The F-22 can carry two GBU-32s internally. (Lockheed Photo)

GBU-31/32 Joint Direct Attack Munition (JDAM)

Manufacturer: Boeing
Native Users: USAF, USN, USMC
Foreign Users: none
Variants: GBU-31/32
Launching Aircraft: AV-8B, B-1, B-2, B-52, F-15E, F-16, F-22, F-117A, F/A-18
Operating Period: 1999-current
Mission: stand off precision strike, all weather capable
The Story: The only true all weather capable bomb currently in the inventory, the JDAM was used operationally for the first time by the B-2 bomber going downtown to Belgrade during Operation ALLIED FORCE in 1999. The B-2 carried 16 JDAMs, and could take out 16 different targets on one pass. The capability to target multiple targets in one pass is a huge advantage the JDAM has. To achieve this capability, the JDAM is the first bomb to use the Global Positioning System in the targeting process. Program in the coordinates, fly near the target, and release. The bomb then flies itself to the target, regardless of the weather. The plane never has to visually acquire the target, like it must to use laser-guided bombs.

The GBU-31 is the 2,000 lb warhead version of the JDAM, and the GBU-32 is the 1,000 lb warhead version

GBU-31/32 JDAMs in testing on an F-15E. Note the afterburners are lit on the Strike Eagle, making this a high speed, high "G" evaluation. (McDonnell Douglas Photo)

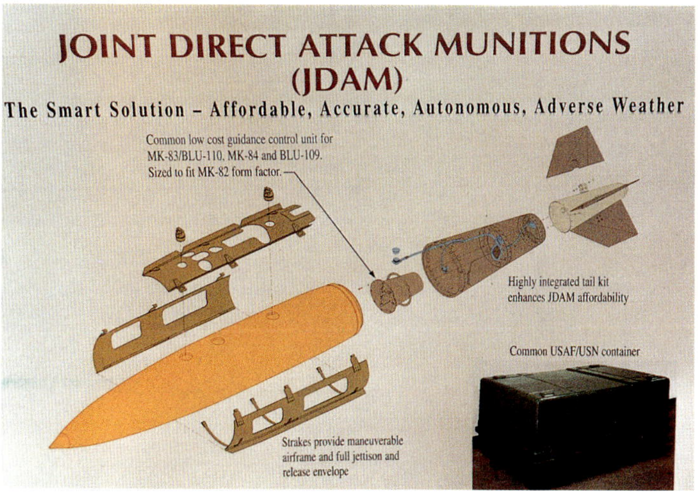

An artist's breakdown of the JDAM sub-assemblies. (McDonnell Douglas Photo)

WCMD
Manufacturer: Lockheed Martin
Native Users: USAF
Foreign Users: none
Variants: none
Launching Aircraft: B-1, B-52, F-15E, F-16, F-117, F-22
Operating Period: 1998-current
Mission: A tail kit that can be added to unguided bombs, giving them precision strike capability
The Story: This smart tail kit can be added to dumb bombs, giving them the ability to become precision strike weapons. The tail kit involves an inertial navigation systems (INS), movable tail fins, and an electronic signal processor. The combination allows the bomb to correct for wind when dropped from high altitudes. Initial testing and development began with the cluster bomb units (CBU) 87, 89, and 97 dispensers. These CBUs can carry mines, cluster bombs, or anti-armor munitions.

A pair of GBU 31/32 Joint Direct Attack Munitions (JDAM) on an F/A-18 during captive-carry tests. (USAF Photo)

A CBU-87 bomb is led onto the wing of an A-10 at Gioia del Colle Air Base, Italy, during Operation Allied Force. (USAF Photo)

Cluster Bomb Units
Manufacturer: ISC, Textron, various others
Native Users: USAF, USN, USMC
Foreign Users: too numerous to list
Variants: CBU-52/-58/-71/-87/-89/-97, Rockeye II, SUU-65
Launching Aircraft: too numerous to list
Operating Period: WWII-present
Mission: various
The Story: Cluster Bomb Units have been used since WWII when the Allies created a bomb that, once dropped, split open and scattered numerous other bomblets on and near targets. These weapons were most effective against soft targets, like unarmored vehicles, troops, fuel and storage depots, etc. The basic process has not changed in the past 50 years, only the munitions themselves. Nowadays, CBUs can carry Fuel-Air Explosives (CBU-55), anti-armor (CBU-97, the Sensor Fuzed Weapon), cluster bombs (CBU-87), and a combination of hard and soft targets (Rockeye II)

Though all these weapons vary in size and type of munitions, they all are basically employed the same way. An aircraft makes a run over the target area, drops its bomb, and the bomb then subdivides into separate munitions. The number of submunitions varies widely. The CBU-55 carries three BLU-73 Fuel-Air Explosives, while the Rockeye II carries 247 bomblets.

Most of the submunitions are dumb, in that they fall from where they are dropped, doing damage in that area. A technological step forward was taken when the CBU-97 went operational. This weapon, once it deploys its submunitions at a preset altitude, then deploys its warheads. Once deployed, an infrared sensor in each warhead looks for both moving or stationary targets on the ground. Once a target is selected, an explosively formed penetrator is fired into the top of the target.

At least one new cluster bomb has been developed for the F-117 and used during the Air War over Serbia. Called the CBU-102, the weapon is a cluster bomb filled with not explosives, but very fine, hair-like fibers that are coated with a conductive coating. These bombs were used on power grids, causing massive short circuits and resulting in power outages to more than half of Serbia. The CBU-102 doesn't destroy the power plant only causes massive short circuits, blowing circuits all over. Serbs could, and did, clean up the mess, but only after many, many hours of effort. The bomb is supposed to carry about 150 spools of the fiber, each spool containing more than 400 feet of the fiber.

B-Series Nuclear bombs
Manufacturer: Various
Native Users: USAF, USN, USMC,
Foreign Users: various NATO allies
Variants: numerous, depending on fuzing options, internal or external carriage, yield size
Launching Aircraft: bombers and tactical
Operating Periods: 1960-current
Mission: Strategic and tactical nuclear bombs
The Story: The ultimate weapons, developed to stop the anticipated USSR invasion of Europe, the Communist domination of the Far East, and the Soviet attack on the United States. A head spinning array of possibilities! We should all be thanking whatever deity we

A B-1B Spirit bomber from South Dakota drops a Sensor Fused Weapon, a high-tech cluster bomb, on the Eglin AFB, FL, test range. Until this test, the only Air Force jets to drop the Sensor Fused Weapon at Eglin were F-16 Falcons. (USAF Photo)

worship that the time to use these weapons never came to pass. Doing the research for these weapons, two things became clear. One, there is not a lot of information, other than the minutia of yields and employment characteristics out there. And two, there were a lot of these bombs built. Just among the ones listed in this entry, more than 10,000 of these nuclear bombs were produced and put into operational service. Granted, many of them have since been retired, but the amount of damage they could have done is staggering.

The yield in these weapons range from "just" 10 kilotons (10,000 tons of dynamite) to nearly 10 megatons (10 million tons of dynamite). They could be used as either airburst, creating an overpressure that would destroy anything within miles of the blast, or ground burst, to destroy buried or hardened command and control bunkers, or other buried/hardened targets.

The Future

Advanced Anti-Radar Guided Missile

Manufacturer: None
Native Users: U.S. Marine Corps
Foreign Users: None
Variants: None
Launching Aircraft: Fixed wing and helicopter
Operating Period: None
Mission: Anti-Radar
The Story: The AARGM is being considered by the U.S. Marine Corps as a follow on to the AGM-122 Sidearm, sometime after the turn of the century. It is designed to be launched from every platform that flies for the Marines. It should have a range of about 10-12 miles, and a dual seeker system (active and passive) is proposed.

Hypervelocity missile

Manufacturer: Boeing
Native Users: USAF, USN
Foreign Users: none
Variants: none
Launching Aircraft: proposed B-1, B-2, F-15E, F/A-18, and F-22
Operating Period: none
Mission: Deep penetrating air-to-surface missile
The Story: Originally begun in 1981, the HVM program developed as a ground launched, surface to surface missile.

In 1998, the Defense Advanced Research Projects Agency gave Boeing a $10 million contract to develop two hypervelocity missile concepts.

One design uses the hypersonic waverider concept. The other concept is the more traditional ramjet powered concept.

Whichever version is picked for further development, it needs to have a range of at least 400 miles and fly at speeds of Mach 6. The GPS/INS guidance system will guide the 250 lb warhead to the target. The warhead can be either a smart submunition, a penetrating warhead, or a standard high explosive/fragmentary warhead.

The size of the warhead may seem small, but combine it with the kinetic energy of impacting the target at 4,000+ miles per hour, and the target will be destroyed.

The Air Force had hoped to have the missile in the active inventory by 2015, but that plan is continuously threatened by the budget cuts and the fact that the HVM is not a high priority program. A flight demonstration/proof of concept flight may take to the air by 2005, if the program stays on track.

The missile, as it now stands, is designed to be used against mobile missile launchers (SCUDs) and hardened, buried targets. When fired, the HVM would climb to 100,000 feet or higher, nose over, and power its way to the target. Reaching speeds of up to Mach 8 in its descent, it would destroy its target. It would have a range of more than 500 miles and use both a radar and laser targeting system.

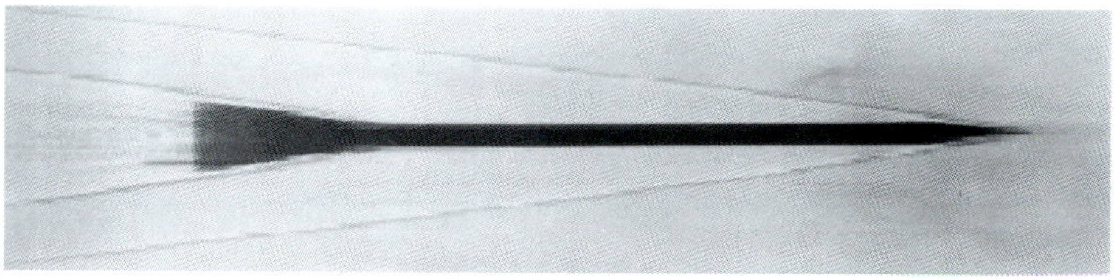

A computer model of what the airflow over a Mach 6 hypersonic missile would look like. (USAF Photo)

5

Foreign Air-to-Ground Weapons

Many military professionals acknowledge that the U.S. is no longer the sole provider of world class weapons. The playing field is darn near even, with the British, French, Russian, and Chinese being leaders in developing and fielding world class weapons. Certainly these countries can be considered among the elite, but do not count out Israel, Japan, South Africa, Norway, Germany, and Sweden as top notch builders.

Probably the biggest difference between these non-U.S. companies and their U.S. counterparts is that many of these companies are not truly private companies, forced by shareholders to return value on their stock. Most of the companies are receiving subsidies from their governments to conduct research and development. Whether or not this is fair is immaterial; the weapons produced by these companies are world-class, and the U.S. has bought some in the past, and will most likely continue to buy in the future.

The next great revolution in weapons will be the growing use of GPS guidance systems. These types of weapons will turn any strike aircraft into a precision attack bomber, and for very low cost and tremendous flexibility. With GPS, any plane can become an all-weather attack plane. GPS guidance systems can be attached to regular iron bombs, short and medium range attack missiles, and long range, stand off missiles. Also, GPS is showing up more and more often as the guidance system of choice for cruise missiles.

A French Mirage 2000N receives fuel from a KC-135 over the Adriatic. The Mirage 2000N was flying a mission supporting NATO peace force operations in Bosnia. The Mirage 2000 can carry a variety of air-to-ground weapons. (USAF Photo)

A German Air Force Panavia Tornado of JbG-32, Lechfeld, Germany, heads back toward its patrol area after refueling from a USAF Europe KC-135R Stratotanker. JbG-32, or Fighter-Bomber Wing Thirty-Two, is one of the German Air Forces' premier combat aircraft wings. The Tornado can carry the Taurus KEPD 350 missile. (USAF Photo)

An Italian Air Force Panavia Tornado is refueled by a USAF KC-135 during Operation Allied Force. (USAF Photo)

Speaking of cruise missiles, as you read this chapter, you will notice a growing trend—more and more companies are producing and selling cruise missiles of various ranges. Cruise missiles are now in the operational inventory of at least 40 third world military forces. This profound growth is explained as a military necessity. Compared to ballistic missiles, which are unbeliveably complicated and expensive, a decent cruise missile can be had for $500,000 or less. And the payoff for third world nations that lack the prestige and operational flexibility of large military establishments is huge.

Cruise missiles offer these countries the ability to defeat a major naval combatant in a superpower's navy. Despite the vast differences in gross national product and military capability between third world nations and the industrialized powers, one accurately placed cruise missile launched from a third world patrol boat or coastal patrol aircraft is capable of achieving strategic results. Argentina's use of Exocet cruise missiles in the Falklands War against the British Royal Navy furnishes perhaps the best example of both how effective cruise missiles can be and how close a third world power came to achieving strategic results with just one weapon system.

The U.S. is the most prolific exporter of cruise missile systems in the form of the Harpoon. It can be launched from ships, submarines, and aircraft; uses a turbojet engine for propulsion; and has an active radar seeker for terminal guidance. Overall, the U.S. has transferred Harpoons to 23 nations, including North Atlantic Treaty Organization allies, the Middle East (Iran included), the Far East, and South America. Taiwan has reverse-engineered the Harpoon into the Hsiung Feng-2 (HF-2), which is reportedly for sale.

As defense companies continue to buy each other out or merge with one another, cross-border ties will become more common. Already there are a number of European companies that do business together, and we can expect U.S. companies to begin partnerships with European or Asian companies in the near future.

As in the previous chapter, rather than try to make one long listing of weapons, running dozens of pages, this section will be broken down by countries. The listing is not intended to be complete; with the proliferation of countries making weapons, and the wide variety of weapons being produced, a complete, all-inclusive list is impossible.

What follows is a review, by country, of the significant weapons in use or development. The list of countries is not all inclusive, since that list would be about 150 countries long, so it is limited to the major players who are actively designing, developing, and manufacturing original weapon designs, not simply re-engineering a current weapons system and mass producing it.

Argentina

Mathogo

A short range, wire guided anti-tank missile developed in the early 1970s, the Mathogo is launched from helicopters. A TV-guided version was developed in the late 1980s. It has a range of about 2 miles, with a warhead of only about six pounds

Martin Pescador

This missile looks a lot like the AGM-12 Bullpup, but it was developed independently of any U.S. participation. It is a short range attack missile, carrying a warhead of nearly 100 lbs. It has a range of about five miles. It can be carried by the A-4, Super Entendard, IA-58, MB-326, and the A109 helicopter. The helicopter version has a range of about three miles.

MP-1000

A wire-guided version of the Martin Pescador, launched exclusively by the A109 helicopter. Has a range of about three miles

China

HJ-8A/B

The HJ-8 is a short range, wire guided anti-tank missile, an upgraded version of the earlier HJ-7. Both are based on the Russian AT-3. The first of these anti-tank missiles were used by infantry against helicopters, but they are now used on attack helicopters, as well as numerous vehicles. It has a fairly short range (about 1.5 miles) and carries a 10 lb warhead.

C-101

A very unique looking medium range, radar guided missile. It is powered by two ramjet engines that can power the missile to nearly Mach 2. At nearly 25 feet long, this is a big missile. It can be air launched by several large aircraft, including the Chinese built versions of the Tu-16 Badger.

HY-4 (C-201)

A long range, radar guided missile used to attack surface ships. It can be carried on aircraft or launched from ships or shore batteries. It is similar in size and shape to the YJ-6 (see next entry), but uses a turbojet motor instead of a liquid propellant. Newer versions have a range approaching 150 miles, with a 1,000 lb warhead.

YJ-6 (C-601/CAS-1 Kraken)

Known by all three designations, this missile is a long range, radar-guided air-to-surface weapon. It is huge, measuring more than 21 feet in length. It was designed in the 1970s to attack large surface ships. To that end, it carries a 1,000 lb warhead and has a range of about 50 miles. The C-601, the export version of the YJ-6, has shown up in Iraq and Iran as the Silkworm shore-launched anti-shipping missile. During Desert Storm, these missiles kept the U.S. Navy on there toes, though no U.S. ships were damaged by them. The Chinese versions can be launched from ships, shore, and the air, primarily from the Chinese-built TU-16 Badger bomber.

YJ-1/C-801

Looking a lot like France's Exocet missile, the C-801 is the export version of the YJ-1. Although primarily a ship and shore launched anti-shipping missile, the Chinese do have an air launched version. It carries a 300 lb warhead with a range of about 25 miles.

YJ-2/C-802

This missile is exactly like the YJ-1, but it is about three feet longer, carries a turbofan engine (not the solid rocket motor on the YJ-1), and more fuel, resulting in increased range—up to 75 miles.

France

AS-11

Basically a small helicopter carried, wire guided missile. Its significance is that the AS-11 traces its roots back to the very beginning of missile development in 1946. The AS-11 was constantly upgraded and finally ended production in the 1980s. The missile can be tipped with five different types of warheads, each about 10 lbs. The warheads include semi armoring piercing, high explosive, fragmentation, and a practice round. It was bought by the U.S. Army, and given the U.S. designation AGM-22. Over the years it was exported to more than a dozen countries, including Iran, Iraq, Libya, UK, and U.S.

AS-12

A short range, wire guided missile with a warhead about six times bigger than the AS-11. This missile was initially developed as a surface to surface missile, but it was quickly adapted to an air to surface role. A short time later, the French Navy made a version to put on their ships. It is carried on both helicopter and fixed wing. The 60 lb warhead can be semi armor piercing, shaped charge, or fragmentation. More than 25 countries bought the AS-12, including Angola, Argentina, Cameroon, Germany, Iran, Iraq, Turkey, and the UK. In a notable engagement, AS-12s, fired from Royal Navy helicopters, damaged an Argentine submarine during the Falkland Island conflict, leading to its eventual capture.

AS-15TT

An all weather, medium range, radar guided anti-ship missile. It carries a 70 lb high explosive warhead. Primarily carried on helicopters, some coastal defense aircraft are also able to carry the AS-15TT. The AS-15TT entered active service in 1985 with the French and has been exported to about ten other countries.

AS-30

A short range, laser guided missile. It is the third generation of its AS-20 and AS-25 predecessors. Though not a full up fire and forget missile, once launched, the launching aircraft can maneuver to stay a safe distance away from the target. Earlier versions required the pilot to keep the target in his cross hairs until impact. The missile can be attached to a pod carrying a TV or IR seeker, allowing the missile to be used at night or in other low light conditions. The missile, with its 500 lb warhead and six mile range, can be carried on the Jaguar, Mirage F1, Mirage 2000, F-16, Rafale, Super Entendard, Tornado, and AMX. The French Air Force fired numerous AS-30s during Desert Storm. It has been exported to about 10 countries, including Egypt, India, Iraq, Pakistan, and Jordan.

ASLP/ASN

In private development, the ASLP/ASN is a long range air to surface missile with a stealthy look to it. It is about 15 feet long and is powered by a solid fuel rocket/ramjet combination. The two will power the missile to speeds in excess of Mach 3. Reports claim the range is in the 500-800 mile range, depending on launch altitude. The missile is designed to be carried by the new French fighter, the Rafale.

AM39 Exocet

One of the most famous and battle tested medium range, radar guided anti-ship missiles in the world. The Exocet was made famous by its devastating impact in the Falklands War. During the opening days of this war, HMS *Sheffield* was attacked by the Argentinian air force on May 4, 1982. The loss of the *Sheffield* was a surprising blow to the British navy, and let the world know that the Exocet was a for-

The Aerospaitial-Marta nuclear tipped cruise missile on a French Navy Super Etendard. (Aerospatiale Matra Photo)

midable weapon. In May 1987, two Exocets, fired from an Iraqi Mirage F1, severely damaged the USS *Stark*, killing [XX] sailors.

The Exocet carries a 300 lb shaped charge warhead and has a range of about 35 miles. An updated version, the block 2 version, has extended the range to about 50 miles.

It carries a 350 lb high explosive, shaped charge fragmentaion warhead.

ARMAT

A medium range, anti radar missile is an offshoot of the AS-37 Martel, but the ARMAT has a better passive radar seeker and electronics. This is strictly a French developed missile, as opposed to the Martel, which was a joint French-British effort. It is carried by the Jaguar, Mirage F1, F3, and 2000 aircraft. It can be launched from low, medium, or high altitude. At high altitude, it has a range of about 50 miles. It carries a 300 lb high explosive warhead.

ASMP

A medium range air-to-ground missile powered by a ramjet with integrated booster.

Carrying a tactical nuclear warhead, this supersonic missile has an extensive flight envelope. It is guided by a highly accurate inertial navigation system that allows the launching aircraft to stay out of range of enemy defenses. Other versions have been developed to carry convential explosive warheads (ASMP-C). An even newer variant, the ASMP-R, carries a passive anti radar seeker and can be launched at airborne electronic warfare aircraft, such as the AWACS, as a long range, supersonic air-to-air missile. The ASMP is carried on the Super Entendard, Mirage 2000N, and Mirage IV. It has range of upwards of 200 miles and a very wide flight envelope—cruising speed in excess of Mach 3 at high altitude and Mach 2+ at low altitude.

A follow on missile, the ASMP-A, is being studied by the manufacturer, Aerospataile. See next entry.

ASMP-A

The ASMP-A will replace the ASMP as the French Air Force's air launched nuclear weapon late this decade.

Though it carries the same name, it will be an all new missile using a similar airframe size in order to minimize the aircraft carriage pylon or ground equipment.

A rocket-ramjet propulsion system is being eyed to power the ASMP-A. The solid rocket motor is the boost engine, and the ramjet is the flight sustainer motor. This combination offers a high velocity missile that is lighter in weight, though more expensive.

The planned missile will have an estimated range in excess of 300 miles. The missile will be carried by the Mirage 2000N and the Rafale.

The upgraded missile allows France to remain the only European country to maintain an air delivered nuclear capability.

Apache

Under development since 1989, the Apache is modular in design and comes in three versions, one of which is the Apache AP—the original 100 mile range version carrying a submunitions payload

A French Air Force Mirage 2000N launches an Apache subsonic cruise missile during test and evaluation. The Apache is a medium range cruise missile for attacking landing strips and highways. (Aerospatiale Matra Photo)

of about 1,100 lbs. This version will be deployed by the French and German air forces and is being offered for export, although no firm contracts have been reported. Initial missile deliveries began in 1997.

Apache AI—an extended-range (175 to 250 miles) version with a 900 lb warhead, was recently approved for series production in 1999 for the French air force and navy. A derivative of the Apache AI also will be offered as a competitive bid in the British conventionally armed standoff missile (CASOM) procurement.

Apache Scalp—the longest-range (250 to 400 miles) version slated exclusively for French use as a nuclear weapon under the control of the General Staff. It is expected to enter series production in 2001.

The Apache family of cruise missiles was designed from the beginning with high terminal effectiveness in mind: they have a stealthy aerodynamic shape and very low-level, terrain-following flight characteristics.

The Apache cruise missile on a French Air Force Mirage 2000N. (Aerospatiale Matra Photo)

All versions of the Apache are powered by the same turbojet engine. Variations in fuel loading account for differences in range among the three versions.

The Apache, when carrying the submunition warhead, has the ability to attack multiple targets on or near landing strips and highways. Once fired, Apache will automatically fly itself to the target. It is an all weather, day/night weapon that carries ten antirunway sub-munitions in its cargo section. It can be carried on all NATO aircraft, including the upcoming Eurofighter, F/A-18, F-16, Mirage 2000, Tornado, and the Rafale.

The British developed their version of SCALP as a stand off missile, called the Storm Shadow.

Durandel

This important runway-cratering weapon was developed in the mid 1970s by Matra for the French Air Force's requirement for an anti-runway weapon which could be delivered in high-speed, low-level flight. The missile had to have a high probability of runway cratering, as well as penetration and destruction of hardened aircraft shelters.

The Durandal can be released at any altitude down to 195 ft, and at any speed from 400 to 640 mph. Upon release, the weapon is slowed by a double-parachute retarder, which diminishes forward velocity to only 56 mph. At this point, the nose of the weapon pitches downward, pointing down toward the runway. When the optimal angle of attack is reached, the parachute is released and the internal rocket motor ignites. The weapon accelerates to over 500 mph before impacting on the runway surface. The weapon's kinetic energy is enough for the 220 lb, hardened steel warhead to penetrate 15 inches of concrete. A one-second delay after impact ensures that the warhead will detonate below the runway's surface.

The Durandal is used by 11 different countries, including the U.S. The U.S. Air Force designates the weapon as BLU-107.

Scalp

Actually a derivative of the Apache, the Scalp is the longest-range (250 to 400 miles) version of the Apache. Initially slated exclusively for the French to use as a nuclear weapon carrier, it may acquire a long range submunition carrier role. The Scalp is expected to enter series production in 2001. For more detailed information on this family of missiles, see the Apache entry.

A Kormoran 2 anti-ship missile in action. (DaimlerChrysler Aerospace Photo)

A Kormoran 1 anti-ship missile being fired from a Luftwaffe F-104G. (DaimlerChrysler Aerospace Photo)

The Panavia Tornado, here in Luftwaffe service, can carry up to four Kormoran 1 anti-ship missiles. (DaimlerChrysler Aerospace Photo)

The Taurus KEPD 350, a stand-off weapon system, is the first European long-range stand-off missile which can be engaged against area and point targets. It was adapted for the Tornado and is being proposed for the Eurofighter. (DaimlerChrysler Aerospace)

Germany

AS.37 Kormoran

A medium range anti-ship missile, the development of the Kormoran began in the early 1960s, based on a request by the German Ministry of Defense Luftwaffe in the early 1960s. The French also participated in the development and testing, but the German company of Messerschmitt-Bolkow-Blohm was the prime contractor. Even though the French participated, they did not purchase the Kormoran. The Kormoran was built until 1983. A follow on, the Kormoran 2, has a range of about 20 miles and carries a semi armor piercing warhead of about 350 lbs. The Kormoran 2, although similar in size, carries a larger warhed and has slightly more range due to the miniturization of the electronics. Its warhead is about 450 lbs and range is about 25 miles.

The Kormoran is carried on the Tornado, F-104, and MiG-29, and should be fitted to the Eurofighter Typhoon.

Taurus KEPD 350

The Tauras KEPD 350 is a standoff, air launched missile used to attack ground and area targets. It can be tipped with either a single high explosive, fragementation warhead, or a submunition to area attacks. It will have a range of more than 200 miles (350 kilometers, hence the 350 designation) and will be carried by the Eurofighter Typhoon and Panavia Tornados. A smaller variant will arm the Swedish JAS 39 Gripen.

It will build on the experience gained from development of the DWS39 submunition. LFK, the missile subsidiary of Daimler Benz Aerospace, who is prime contractor, will seek to have significant commonality with the Scalp and Storm Shadow cruise missiles being bought by France and the United Kingdom.

A Luftwaffe Tornado fires a Kormoran 1 from a height of about one hundred feet. (DaimlerChrysler Aerospace Photo)

First flight of the Taurus KEPD 350 on a Tornado. (Daimler Chrysler Aerospace Photo)

Israel

Popeye

A medium range TV or IIR guided missile, the Popeye is the result of a joint venture company being established between Lockheed Martin and Rafael. The missile, in the U.S., is designated the AGM-142.

Popeye is a standoff, air-to-ground, guided missile fitted with either a television (TV) or imaging infrared (IIR) seeker. The system also includes an INS and data link pod (DLP).

The missile weighs 3,000 lbs, is powered by a single stage solid rocket motor, and armed with either a general purpose 750 lb fragmentation warhead or an 800 lb penetrating warhead.

The Popeye 2 uses lighter internal electronics and a smaller rocket engine to reduce the weight to the 2,000 lb range, allowing the IAF F-16s to carry two of these missiles into combat.

The Popeye 3 is turbojet powered and has a range of about 200 miles.

NIMROD

NIMROD is a laser guided anti-tank missile with a range of about 15 miles and pinpoint accuracy. At around 10 feet long, it is larger than most anti tank missiles, but it aslo has a longer range than most. It is easily installed on a variety of ground platforms and helicopters in sealed 4-missile canister units.

NIMROD is capable of day/night simultaneous multiple launches against several targets using several designators. Its altimeter-set flight trajectory allows flight below clouds, adjusting to varying weather conditions.

Gabriel 3AS/4LR

A medium and long range radar guided, anti ship missile. Earlier versions (Gabriel 1 and 2) began development in the 1960s and saw operational use during the Yom Kippur war in 1973. These two newer versions are very similar to its predecessors, but with a slightly lengthened body.

The 3AS can target a variety of ships, from fast attack to destroyers. The 3AS has a range of about 20 miles, and a 300 lb warhead. It can be carried on all of the Israeli aircraft.

A long range version of the 3AS, dubbed the 4LR, increases range to about 150 miles, and carries a 500 lb warhead.

Star 1

The Star 1 began as the Delilah, a missile decoy, but it was adapted to include offensive capability.

The Star 1 is an anti radar version of the Delilah. It has a GPS midcourse guidance system and a passive seeker in the terminal phase. It weighs about 500 lbs, and has a range of approximately 60 miles.

Another adaptation of the Delilah decoy is the Delilah cruise missile.

Italy

Marte

Numerous versions of this medium range, radar guided anti ship missile were built. The Marte 1 began development in the late 1960s, then known as the Sea Killer. The next version, the Marte 2, began life in 1980 with a number of improvements over the earlier model. Among the changes was an upgraded radar.

The Marte is one of the most unique looking missiles, with a large bulbous nose that narrows to a long, narrow frame. It comes in fixed wing, rotary wing, and sea launched versions. The wing

Cutaway drawing of the Taurus Missile System. (DaimlerChrysler Aerospace Photo)

An artist's rendering of the round-a-bout routing the Taurus could take if actually employed. (DiamlerChrysler Aerospace Photo)

fixed version, the 2A, has a range of about 20 miles. The rotary wing version's range is about 12 miles.

The Italian armed forces began deploying the Marte 2 in the late 1980s. The fixed wing version carries a 550 lb warhead, while the helicopter launched version carries a 150 lb warhead.

Skyshark

A proposed missile that was developed, but never saw operational service, the Skyshark is a powered submunition that comes in two versions—a medium range and a long range version. The short range version, about 30 miles, is powered by a solid fuel rocket motor, and the long range version, with a range of about 150 miles, is powered by a turbojet.

The munitions capability varies, too. The short range version carries about 1,500 lbs of sumunitions, while the long range version carries about 2,000 lbs.

Guidance comes from an INS/GPS system that is highly accurate.

Japan

ASM-1

This was the first air to surface missile developed by Japan for its Self Defense Forces. The development began in 1973. It uses internal guidance for the flight to mid course and then activates an active radar for terminal flight. There are two operational versions of the ASM-1, the -1 (Type 80) and -1C (Type 91).

The ASM-1A carries a 300 lb warhead and has a maximum range of about 35 miles. It has been cleared to fly on the F-1, F-4, and P-3s. The -1C boasts an improved rocket motor that makes the missile lighter, thereby increasing its range to about 40 miles.

ASM-2

The ASM-2 is the successor for the ASM-1s currently in service. The ASM-2 is long range. It has improved electronics, a passive radar for the terminal phase, and uses composite material in the missile body and wings to increase its stealth characteristics.

The ASM-2 (Type 93) began development in the late 1980s and entered service in the late 1990s. The missile has a range of about 70 miles.

Norway

Penguin 2/3

The Penguin Mk 3 (U.S. designation is AGM-119) is a Norwegian-designed anti ship missile, capable against small surface ships and surfaced submarines. The IR seeker head is effective against a wide range of targets, and its profile is hard to defend against. It is a short-to-medium range inertially guided missile and is capable of gravity drop launches at low speeds and altitudes. A fire-and-forget missile, the Penguin has a 360 degree arc, autonomous search, acquisition and track during terminal phase. This supersonic missile carries a 265 lb warhead.

The AGM-119A is designed to be used by aircraft and has a range of 30 miles. The AGM-119B is designed for helicopter use and has a range of 18 miles. The missile is currently employed by many nations, including the USAFD, USN, Greece, Singapore, and Turkey.

Russia

AS-4/Kh-22

NATO name Kitchen (all air to surface missiles are given NATO designators that begin with the letter K), this is a long range, radar guided missile for attacking ships and land based targets.

The Kitchen had at least three versions built, all seeing operational service. The differences between the three were in the area of various guidance systems. One was inertial midcourse and active radar in the terminal phase, another had a passive radar guidance system, and a third had active, but carried a nuclear warhead.

The conventionally armed Kitchen carried a 2,000 lb warhead about 250 miles. Some of these missiles are still thought to be on active service, carried on Tu-95 Bear and Tu-22 Backfire bombers.

AS-6/Kh-26

NATO name is Kingfish. This long range radar guided missile targets ships or land based targets, primarily radar systems. It can carry a 2,000 lb conventional high explosive warhead or a nuclear warhead. It is very similar in design to the AS-4, but has a shorter range—about 175 miles or so.

It also has various methods of guidance, from inertial midcourse and active radar to passive radar.

The Kingfish is carried on Tu-16 Badger and Tu-22 Backfire bombers.

AS-7/Kh-23

NATO name Kerry, the AS-7 is a first generation air to surface missile. It was developed in the early 1960s from the Alkali, an air to air missile. The AS-7 is the basis for further developments, including the AS-10 and AS-12.

The missile is radio guided by the pilot using a joystick after lining up the missile with the target. The joystick allows for fine tuning the aim. It carries a 225 lb warhead and has a range of about 3 miles.

The AS-7 was carried on fighter-type aircraft, including the MiG-21, MiG-27, Su-17, Su-22, Su-24, and others. It was exported to more than 30 countries.

AS-9/Kh-28

NATO name is Kyle. The AS-9 is a short range anti radar missile. Instead of attacking the large, land based radar systems that the AS-4 and AS-6 would attack, the AS-9 is designed to take out the radar systems normally associated with surface to air missile sites.

Developed in the late 1960s, the AS-9 is a large missile—almost 20 feet long—and has a range of about 50 miles. It uses a passive radar to lock on and follow a radar beam. It carries a 325 lb warhead.

It was exported to a number of countries, including the usual suspects—former Iron Curtain puppet states, Iraq, Iran, Libya, Vietnam, and others.

AS-10/Kh-25

NATO name is Karen. The AS-10 was developed in the early 1970s as a short range, radio, or laser guided attack missile.

Built in two versions, both share a common 200 lb warhead. One version is radio guided, while the other is laser guided (by the launching aircraft). It has a range of about 6 miles in the radio controlled version and about 12 in the laser guided version.

The AS-10 is carried on a number of fighter and attack aircraft, including the MiG-27, Su-17, Su-20, Su-22, Su-24, and Su-25. Exported to a large number of countries, including former communist countries, Afghanistan, Syria, and Vietnam.

AS-11/Kh-58

NATO name is Kilter. The AS-11 is a medium range, anti radar missile. It was developed in the 1970s as the third generation of Soviet anti radar missiles. Considerably smaller than its predecessors, the Kingfish and Kitchen, the AS-11 was designed to fit on the then front line Soviet fighters, like the MiG-27 and Su-24. Previous anti radar missiles were too big for fighters, and were carried on bombers. As newer fighters came on line, they also were fitted for the AS-11.

It carries a passive radar guidance system and is powered by a solid rocket motor that can accelerate the missile up to Mach 4. Launched from high altitudes, it has a maximum range of over 100 miles.

A variant of the AS-11 is designed to attack ship radars. It carries a conventional, 330 lb high explosive fragmentation warhead. A nuclear warhead has been designed to fit on the missile, as well.

AS-12/Kh-25MP

NATO codename is Kegler. The AS-12 is also a medium range, anti radar missile. It is very similar to the AS-10. It carries a passive radar guidance system and a 200 lb high explosive, fragementation warhead.

The AS-12 has a range of about 40 miles when launched from high altitudes and about 15 miles from low altitudes. Its solid rocket motor pushes the missile to about Mach 2.5.

The AS-12 is carried on most of the Russian front line fighters, as well as most of the aircraft they exported over the past 20 years: MiG-21, Mig-23, MiG-27, Su-17, Su-20, Su-22, Su-24, Su-25, Su-35, Su-37, Tu-22 bombers, and the Ka-50 attack helicopter.

The AS-12 entered active service around 1980 and flies with the Russian, Czech, German, Hungarian, and Polish air forces.

AS-13/Kh-58

NATO codename Kingbolt, this missile is a TV guided, medium range missile. It is powered by two solid fuel rockets firing in sequence—one to get the missile launched, then the second kicks in and powers the missile to the target. With the two engines, the range is 100 miles when launched from high altitude and about 60 miles from low altitude.

The Kingbolt is a big missile, weighing in at almost 1,900 lbs with a 330 lb warhead. It is about 17 feet long. It is carried on the Su-12, Su-22, Su-24, Su-27, and MiG-27.

AS-14/Kh-29

NATO codename is Kedge. The AS-14 is a medium range, laser or TV guided missile. The Kedge's mission and design is similar to the AGM-65 Maverick. Designed in the late 1970s, the Kedge is carried on MiG-27, MiG-29, Su-17, Su-22, Su-24, Su-25, Su-35/37, and Mirage F1 fighters.

It is the most common air-to-ground guided missile in the Russian inventory, and is used for destroying hardened targets, such as reinforced concrete shelters, concrete runways, major railway and highway bridges, and sea-based targets, such as ships and surfaced submarines

The AS-14 carries a large warhead, about 700 lbs, and is powered by a solid rocket motor. It has a range of about 8 miles for the TV guided version and about 6 miles for the laser guided version.

Developed for the Soviet Air Force in the 1970s, the Russian Air Force continues to fly with the AS-14. It has been sold to a number of countries, including Bulgaria, Czech Republic, Germany, Hungary, Iraq, and Poland.

AS-15/Kh-55

NATO name is Kent. The AS-15 is a long range, nuclear armed, air launched cruise missile. Developed in the late 1970s and deployed throughout the Soviet Union, the AS-15 is similar to the U.S. Tomahawk cruise missile.

Using INS and terrain matching guidance systems, the AS-15 could carry a 200 kiloton nuclear warhead about 2,000 miles. The missile is carried by the Tu-95 and the Tu-160 bombers on external racks and internal rotary launchers.

The Ukraine signed a contract in 1998 with the U.S. to destroy more than 1,000 AS-15 missiles in their inventory.

AS-16/Kh-15

NATO codename is Kickback. The AS-16 is a long range, anti ship or anti radar air-to-surface missile. It can carry both a 330 lb conventional warhead or a 350 kiloton nuclear warhead. It is powered by a solid fuel rocket motor and has a range of about 100 miles. Its mission is similar to the AGM-69 SRAM—to blast open a path for the bombers carrying the long range cruise missiles. A later version was developed with the conventional warhead, and can be used against ship targets.

The AS-16 is launched from bombers, either the Tu-22, Tu-95, or Tu-160. The bombers can carry up to 24 missiles internally and externally.

AS-17/Kh-31

NATO codename is Krypton. The AS-17 comes in two versions; one is a medium range anti radar missile with a passive radar seeker, and the second is a long range, active radar seeker missile used to attack shipborne radars like the AEGIS system.

The AS-17 has a unique design incorporating a solid fuel rocket motor with ramjets. The intakes for the ramjets are arrayed around the fuselage of the missile, each about seven feet long. Each ramjet duct has a control surface, so the four intakes also provide the control surfaces for the missile.

The two motors combine to push the missile to a max speed of about Mach 3. Tipped with a 200 lb warhead, the AS-17 has a range of about 120 miles.

The AS-17 is carried on MiG-21, MiG-27, MiG-29, Su-24, Su-25, Su-27, Su-32, and Su-35/37.

At least one country has bought the AS-17 from the Russians—the Chinese.

AS-18/Kh-59

NATO code name is Kazoo. The AS-18 is a long range, TV guided air-to-surface missile. This missile was developed in the early 1990s. With a conventional 700 lb warhead, or 615 lb cluster submunitions, the AS-18 has a range of about 60 miles.

It is powered by a "strap on" turbojet engine. At launch, the missile weighs about 2,000 lbs.

It is carried on the new generation of Russian fighters, the Su-35/37, though it can also be fitted on older ones like the MiG-27, Su-17, Su-22, Su-24, and Su-25.

AS-19/BL-10

NATO code name is Koala. The AS-19 is a long range, nuclear armed, air launched cruise missile. First seen by the west in 1988, the AS-19 began development in the early 1980s.

A large missile, the AS-19 is carried on Tu-95 and Tu-160 bombers. The missile flight profile, after launch, is to fly at altitudes of up to 75,000 feet at speeds of around Mach 3. With this combination, it would have a steep dive at the terminal phase, thereby adding enough kinetic energy to attack deeply buried or reinforced targets.

AS-20/Kh-35

NATO designation is Kayak. The AS-20 began life in 1983 as the SS-N-25 surface to surface. In 1987 an air launched version was developed. Very similar to the AGM-84, the AS-20 is often referred to as the "Harpoonski."

The As-20 carries a 300 lb high explosive, fragmentation warhead. When launched from high altitude, it has a maximum range of about 70 miles.

It can be carried on a wide variety of aircraft, including the MiG-29, Su-24, Su-27, and Tu-142 aircraft, as well as the Ka-27 and Ka-50 helicopters.

Kh-41 Moskit

The Kh-41 lacks a NATO designation, but still much is known about this missile. Similar to the AS-17/Kh-31, it is also powered by a rocket/ramjet combination. The missile is a long range, INS, and radar guided anti ship missile.

This huge missile carries a 700 lb high explosive, fragmentation warhead. It uses the INS guidance to fly low levels towards the target area, and then the radar, which can be either active or passive, is used in the terminal phase of attack.

Weighing in at almost 10,000 lbs, it is too big to be carried by most fighters. Only the Su-27 coastal defense fighters can carry this missile up and launch it. It is too heavy for carrier launched Su-27s to launch.

It has a range of about 175 miles when launched from high altitudes.

KH-65SE

Not much information is available on the Kh-65, but it is basically a reduced range AS-15 Kent cruise missile that is sold for export.

KH-101

Also not much information, other than it is believed to be the newest generation of stealthy cruise missiles, carrying either a conventional or nuclear warhead.

KH-SD

A planned replacement of the AS-15 Kent cruise missiles in the Russian inventory.

AFM-L

A future anti ship missile, but lacking funds to begin development.

ALFA

A supersonic anti ship missile being developed without state funding. Most likely, it will be carried by Su-32 coastal defense fighters. It is about 20 feet long, carries a 900 lb warhead, and has a range of nearly 400 miles. It uses INS for midcourse guidance and active radar for the terminal phase.

South Africa

Torgos

A growing producer of air-launched weapons, the South African company Kentron is entering the cruise missile field with an excellent entry called Torgos.

Torgos is a long range missile that can carry either a high explosive or a submunition dispenser. Currently in development, the Torgos is claimed to have a range of about 225 miles. Guidance comes from a GPS/INS system that the company claims will deliver the missile to within six feet of the target.

Despite weighing in at around 2,000 lbs, the Torgos can be carried on the single engine F-16 and the Mirage 2000, two of the most popular fighters in the world. It will also fit on the JAS 39 Gripen and the Hawk, a training aircraft with light combat capability.

Spain

Alada

A family of unpowered and powered weapons, the ALADA is designed for the Eurofighter (the Typhoon), the EF-18, and Harriers flown by the Spanish Air Force and Navy.

The turbojet will power the powered version and will be able to carry either submunitions or a 1,000 lb high explosive warhead.

Sweden

RB04

The Robot 04 family of missile were originally developed in the mid-1950s, and encompassed 5 variants—the RB-4A/B/C/D/E. All versions were medium range anti ship missiles, and had intertial mid course guidance with active radar in the terminal phase. They had a range of around 20 miles

The RB04 family of missiles are retired from active service.

RB05

Also now retired, the RB05 was short range, command guided missile used to attack heavily defended land targets, ships, and even some aircraft, if they were moving slow enough.

Originally, the guidance system required the pilot to launch the missile and basically follow it to impact, keeping a flare in the tail aligned with the target via a datalink. Some testing was done to try and fit a TV guidance system into the weapon, but it never made it to operational service. It had a range of about 5 miles.

RBS 15F

The follow on missile to the RB04, the RB15 is a family of anti ship missiles that can be carried on aircraft, land vehicles, or ships. The development of the air launched version began in the early 1980s.

This family of missiles uses an inertial guidance system for mid course and active radar for the terminal phase. It carries a 450 lb warhead. From high altitude, it has a range in excess of 60 miles. A newer version is thought to have a range of around 150 miles

United Kingdom

HAKIM (PGM-1/2/3)

The product of a convoluted history, the Hakim is a family of medium range, laser, or IIR guided missiles. The Hakim began development by an American company for a middle eastern customer. The company was bought by another, which in turn was bought by GEC-Marconi, a British company.

Development began in the mid 1980s, and it finally saw operational service in the mid 1990s by the United Arab Emirates.

The family of missiles is basically one of two missiles with two different sized warheads and two different sized solid fuel rocket motors. A variety of guidance systems can also be used.

The PGM-1s carry a laser designator, while the PGM-2s carry a TV guidance system and the PGM-3s use an IR seeker.

Each missile also comes in either an A or B version. The A versions all have a constant diameter body, and use the standard NATO attachment lugs. The B versions have a wider body that tapers down to the standard diameter at the warhead. The B versions also have two rocket motors instead of the one on the A versions.

Warheads vary from 2,000 lb penetrating warhead, high explosive/fragmentation, anti tank, and bomblet submunitions.

The range of all versions, whatever the combination, is about 15 miles.

Sea Eagle

The Sea Eagle is a computer-controlled, fire and forget, sea skimming anti-ship missile. It has an active radar target seeker, and its long range allows a covert approach by the launching aircraft. After launch, the missile is autonomous, its trajectory and homing being controlled by on-board computers. The Sea Eagle began to enter RAF service in 1986, the structure and aerodynamic basis for the weapon being the Martel air-to-surface missile. The Buccaneer W2B was the first RAF aircraft to carry Sea Eagle, four of which were carried on wing pylons. Following the Buccaneer's retirement in 1993, the Tornado GR1B is now the RAF's launch platform for this missile.

Storm Shadow is a long-range, stand-off attack missile which combines pinpoint accuracy with very high levels of survivability and warhead effectiveness. (RAF Photo)

Sea Eagle is a computer-controlled, fire and forget, sea skimming anti-ship missile. (RAF Photo)

Stingray

This weapon is carried by the Nimrod in its capacious bomb bay. Stingray is a homing torpedo, and one of the fastest weapons in its class in the world. It is fully programmable with a number of search options, and is designed specifically for tracking and destroying submarines. The seeker head of the torpedo is able to home on to submarine targets either passively, by following the noise the submarine makes, or actively, by pinging with its own sonar and following the sound energy reflected by the target.

Sea Skua

Successfully used in the 1982 Falklands war, the Sea Skua did not officially enter operational service until the next year. Also used during the Gulf War, the Sea Skua is said to have sank more than a dozen Iraqi ships.

The Sea Skua is a medium range radar guided all weather anti ship missile that carries a 50 lb warhead and has a range of about 12 miles.

Development began in the mid 1970s, and it was pulled from evaluation and testing to go to war in the Falklands. Designed to attack primarily small surface ships from helicopters, it proved to be very successful.

A radio altimeter is used for altitude management, while a semi active radar is used to track and attack at target. It has been sold to, among others, Germany, Turkey, and South Korea.

Sabre

A developmental system anti tank missile from the late 1970s that was based on the Rapier surface to air system, the Sabre was a lightweight air to surface missile. In its design, it used a laser designator set for low altitude, all weather use.

Storm Shadow

Another new weapon in the RAF's arsenal, Storm Shadow will form the cornerstone of the RAF's ground attack capability well into the 21st century. The British developed version of the Apache Scalp built by British Aerospace, Storm Shadow is a long-range (250 to 400 miles), stand-off attack missile which combines pinpoint accuracy with very high levels of survivability and warhead effectiveness. It can carry either a conventional warhead, submunitions, or nuclear warheads. The Storm Shadow has a stealthy aerodynamic shape and very low-level, terrain-following flight characteristics. It is powered by a turbojet engine. It is an all weather, day/night weapon that can be carried on all NATO aircraft, including the upcoming Eurofighter Typhoon (EF2000), F/A-18, F-16, Mirage 2000, Tornado, Harrier, and the Rafale. A fully autonomous weapon, Storm Shadow will allow the RAF to attack high priority targets deep within enemy territory, while minimizing the risk to the attacking aircraft. For more information, see Apache entry in the FRANCE section.

ALARM

The ALARM (Air-Launched Anti-Radar Missile) is designed to destroy ground-based air defense radars and surface-to-air missile radars, thereby providing support to attack aircraft while penetrating hostile air defenses. It does this by homing on to radar transmissions and following them down to their source.

It entered service in the early 1990s and proved very effective in the Gulf War. A Tornado GR1 can carry seven ALARMs. One unique feature is the optional capability to loiter over a target area while continuing to search for targets. It provides attack aircraft with an on-board suppression capability which will enable them to press home attacks against heavily defended targets with a high probability of success.

Brimstone

The BL755 will be replaced in the near future by this new missile built by GEC-Marconi. The Brimstone missile is a development of the well proven Hellfire system, and will provide a lethal anti-armor capability for RAF fixed-wing aircraft. Brimstone is autonomous with a stand-off capability that will minimize the risk to the attacking aircraft. Each weapon pylon load will consist of three missiles on a triple launcher, allowing an attack aircraft to carry a great many of the weapons.

BL755

The BL755 Cluster Bomb is carried by offensive support and attack aircraft. Weighing 600 lbs, each bomb carries 147 small

The ALARM (Air-Launched Anti-Armor Missile) is designed to destroy ground-based air defense radar and surface-to-air missile radars. (RAF Photo)

The Brimstone missile is a development of the well-proven Hellfire system, and will provide a lethal anti-armor capability for RAF fixed-wing aircraft. (RAF Photo)

bomblets which scatter over an area on release. A dual-role weapon, it entered RAF service in 1972. It is designed to provide a high "kill" probability against a range of hard and soft targets. When the BL755 is dropped, the primary strike is armed. After a set interval the initial cartridge is fired, which produces the gas pressure to blow off the two-part body skins and then fires the main cartridge. Each bomblet is 5.87" (0.15 m) as it is fired. This extends to 24.69" (0.62m) complete with extending nose probe and retarding parachute.

JP233 Airfield Attack Weapon

A UK-developed submunition dispenser, the JP233 is designed as an air delivered, counter air weapon — it is used to attack an enemy's runways, cratering them to the point where the runway cannot be used.

Development began in the 1970s, with the US a full partner. In 1980 the US withdrew, as they had second thoughts on the method the JP233 would use to attack the airfields. The UK continued with the development, and used the JP233 successfully in the Gulf War. However, the success came at an extremely high cost. The JP233 is designed to be delivered by a high speed, low level attack, flying right down the runway centerline. An extremely hazardous flight profile, as the RAF Tornado pilots found out as a number of them were shot down.

A second version of the JP233 is an area denial weapon, that scatters more than 200 small mines across an area several hundred meters long and wide (exact dimensions depends on drop altitude). The mines can be set to go off immediately, or delay up to 24 hours, a sensor fuze can be attached and the mine will go off when the fuze parameters are met.

Multinational

HOT

A short range, wire guided missile, used primarily by attack helicopters in the anti tank warfare arena. Development began in 1972 under a joint German/French effort to arm both helicopters and ground vehicles with an anti armor capability. The newest version, the HOT 3, is operational, but the Trigat has begun to replace the HOT in both the German and French arsenals. The HOT 1 and HOT 2 missile systems have been sold to more than 25 other nations.

The new HOT 3 began production in 1997 with considerably upgraded kill potential and anti-jamming capabilities. It basically doubles the explosive charge of the warhead over that of previous versions. The HOT 3 serves in the armed forces of 17 nations.

AS 37 MARTEL

A joint UK/French, TV guided anti ship missile. Secondary mission is stand off attack of heavily defended land targets. The missile also has an anti radar seeking capability, to home in on sea or land based radars and follow the radar beam back to the source. The two countries began development in 1964, and the MARTEL entered service in 1970. The MARTEL is still used by export countries, but it has been replaced in both the UK and French inventories by the Sea Eagle (UK) and the ARMAT (French).

The MARTEL carried a 300 lb warhead and had a range of about 30 miles.

Trigat

Third-generation, long-range antitank missile for helicopter and vehicle use. A "fire and forget" type missile using infrared imaging, advanced military technology, and diving-attack principle. It is being developed by a multi-national group involving five nations: UK, France, Germany, Belgium, and the Netherlands.

Trigat will have a range in excess of 5 miles and will be the first anti tank missile to use passive radar, which makes it virtually unjammable. It will lock onto its target before launch, ensuring surgically precise attacks.

Storm Scalp

A conventionally armed stand-off air-to-ground long-range missile. It was selected by the UK to equip the RAF Tornado, Harrier, and Eurofighter, and by the French Air Force to equip the Mirage 2000 and Rafale. It is a fire and forget, all weather, day/night missile that can be used against heavily defended, hardened, and high value fixed targets.

The Storm Scalp has high survivability potential due to the missile's low level terrain following flight and built in stealth and autonomous navigation

It has a range of about 150 miles and uses a specially designed turbojet for propulsion, and INS, GPS, and terrain referencing for navigation.

A HOT launch test firing. HOT is a multi-national anti-tank missile.

6

Other Air-Launched Systems

Introduction

A major portion of the air-launched area is, of course, the subject of the weapons that can be launched from a high-flying platform. But there are other applications of the aerial platform that have been tried in the past. This chapter will address those varied, and oft-times interesting, applications.

In the test area, the technique has proven to be highly effective. A number of significant activities in that arena come to mind. Included are the use of airborne platforms for the launching of unmanned drones for missile intercept missions. There have also been a number of air-launchings of a number of X vehicles, missiles, and aerodynamic vehicles.

Also, there have been a number of programs where an aircraft is carried aloft by both another aircraft, and even a balloon on several occasions, and then launched for its particular mission. With several of these concepts, there was also the capability to recover the craft back to the launching vehicle. Even more bizzare was the possibility of carrying an ICBM aloft and launching it at altitude.

Then, too, there have been air-launched diversionary decoys that have been used to simulate the launching aircraft and confuse enemy radars.

Air-Launch Test Activities

Probably the most memorable was the early flight testing of the Space Shuttle Orbiter vehicle. The technique involved mounting the shuttle atop its Boeing 747 carrier aircraft. This position is, of course, familiar in that it is the established mode of transporting the orbiters back to Cape Kennedy when landings take place at Edwards Air Force Base, Florida.

But this carrier aircraft also served in an entirely different capacity, as that of an air-launch vehicle. The technique involved carrying the shuttle orbiter to altitude and releasing it—with a manned crew aboard—so that an unpowered landing could be accomplished. Recall that there is no landing propulsion contained in the Orbiter, therefore requiring a deadstick landing to be made.

This series of tests in the 1980s proved the techique with a series of highly-successful tests. In fact, it should be noted that there has never been a problem with the many operational landings made through the years.

But the ultimate carrier aircraft is the NASA NB-52 mother ship which first flew in 1953. It is still flying almost five decades later and has been involved in a long series of aircraft, missile, and space test launches.

The NASA Boeing 747 carries the Space Shuttle Orbiter aloft for an unpowered air launch. (NASA Photo)

Air launch of the Space Shuttle Orbiter. (NASA Photo)

X-24 air launch from B-52 mother ship. (USAF Photo)

Among others, the plane has launched the X-24, X-15, and HL-10 space systems, along with serving as the airborne launch platform for the commercially developed Pegasus space booster system. The NB-52 also launched a three-eighths scale model of the F-15 fighter for spin testing. The venerable Stratofortress was initially modified with a pylon located between the fuselage and inboard engine on the right wing. All the air-launched vehicles have been launched from that location.

More recently, there have been a number of new X vehicles launched by the NB-52. The X-34, developed by Orbital Sciences, could be a reusable, air-launched vehicle that could approach speeds of Mach 8. The landing would be a conventional horizontal touchdown.

The X-38 is a vehicle to prove the concept for a space crew return vehicle. The X-38 will be launched by the NB-52, and a soft landing would then be accomplished by a parafoil parachute.

The X-43 Hyper-X vehicle is a test vehicle for future space launch vehicles. The X-43 is mounted on a Pegasus booster rocket, which in turn is mounted on the NB-52, as a sort of piggy-back air-launch technique. Launch will occur between 19,000 and 43,000 feet depending on the particular mission.

Another X vehicle also used an air-launch during its test program. The X-40, a 90 percent scale model of the Space Maneuver Vehicle (SMV), will eventually be launched by a Delta rocket. But during its test program in the late 1990s, the model was lifted into the air by a Blackhawk helicopter, then was dropped, from which it glided in for a landing.

A somewhat bizzare air-launch technique was considered by NASA in the early 1990s for space research. The "launch vehicle" in this situation was an SR-71A, with the launched vehicle being a Mach 10 reusable piloted aircraft carried on a pylon atop the SR-71 fuselage. The plan called for the vehicle to be launched at Mach 3 and 70,000 feet. Called the HALO, the craft had the goal of reaching Mach 9 and 145,000 feet altitude, which was to be accomplished by the craft's liquid hydrogen/liquid rocket motor. The high-Mach vehicle, though, would never fly.

Air-Launched Target Drones

With the advent of the jet age in the 1950s, the need arose for high-performance target drones for the proving of new intercept techniques, and later to serve as targets for the new air-to-air missiles coming on line. Although many such drones were launched from ground launchers, there were a number that were carried aloft under aircraft wings and launched for their missions.

Of course, if the mission was a success, there was a good chance that the drone would be destroyed, and therefore not returning. However, should the drone have survived, many times it was lowered to the ground by an onboard parachute system.

During the 1950s, a number of air-launched unmanned drones were developed. The Northrop XQ-4 supersonic target drone was designed to test the capabilities of new weapon systems. The XQ-4

X-24 carried aloft under wing of B-52 mother ship. (USAF Photo)

X-15 launch to the outer edges of space from the B-52 mother ship. (USAF Photo)

was powered by an underslung Westinghouse J81 turbojet engine, had a radar-controlled guidance system, and an altitude ceiling of over 60,000 feet.

The Northrop XKD4R-1, which was designed to present the ultimate target for air-to-air missiles, was rocket-powered by a solid propellant system capable of driving the missile to Mach 0.9 at 50,000 feet.

During the late 1950s, the Northrop AQM-38 target drone was developed and provided an advanced target for surface-to-air and air-to-air missiles. The AQM-38 was a simple low-cost lightweight target launched from jet fighters and radio-controlled from the ground. It was deployed against the Nike family and Hawk missiles with high performance at high and low altitudes.

The system's solid propellant rocket engine was the main structural component. Aluminum honeycomb wings, plastic nose section, and aft fuselage were attached to the steel engine case. The flight control package, including control vanes, was located in the nose section. Recovery after flight was accomplished by a two-stage parachute.

Another early drone was the Temco XKDT-1 expendable target drone capable of reaching Mach .95 at 50,000 feet. The drone had a dual mission of training jet pilots and for missile intercept capabilities.

The early 1960s saw the introduction of the Northrop RP-76-4 target drone, along with numerous others. But the air-launched drone that got the most attention for many years was the Ryan Firebee system.

The Firebee was initially designated the Q-2, which served as a target for both surface-to-air and air-to-air missile system testing for both the Air Force and Navy. Flexible in its use, the Firebee could be either launched from a ground launcher or off the wing of an aircraft. In either launch environment, the Firebee was controlled from the ground. Once hit by a missile, or completing its mission undamaged, the Firebee's self-contained parachute lowered it to the ground.

The Firebee weighed in at 1,800 pounds and was powered by a Continental J69 turbojet engine with one thousand pounds of thrust. It was 17 feet long, had a wing span of 11 feet, two inches, and contained electronic equipment to score miss distances automatically.

With its effectiveness in a test environment, the Firebee would also find its way into a number of operational missions. The AQM-

An X-34 air-launched space vehicle. (NASA Photo)

Artist's concept of the NASA X-43 space vehicle. (NASA Photo)

The air launched Radioplane XQ-4 drone had a 60,000 foot maximum altitude capability. (USAF Photo)

The Temco XKDT-1 was a rocket-powered, air-launched target drone. (U.S. Navy Photo)

34L Firebee served as a reconnaissance drone which was used for surveillance missions during the Vietnam War. The missions involved air launch from a DC-130 director aircraft which controlled its flight. Following its photographic mission, the Firebee was diverted to a safe area where its parachute was deployed and recovery made for another mission. On some occasions there were air-recovery missions accomplished.

During more than a decade starting in 1964, more than one thousand Firebees flew over 34,000 missions. It is hard to imagine the number of lives saved had those missions been accomplished by manned aircraft.

In recent years, there have been a number of improvements made to U.S. target drones. The BMQ-74E drones have incorporated an Integrated Avionics Unit (IAU). The new system eliminated the need for many existing subsystems which allowed a payload increase, also allowing volume for a Global Positioning System receiver for more accurate navigation. The 13-foot system has the capability to simulate both high-performance fighters and cruise missiles.

Early XAAM-1 Firebird drone. (Air Force Museum Photo)

The Radioplane RP-70 (XKD4R-1) missile target. (U.S. Navy Photo)

A Firebee target missile carries teeth on its nose. (Air Force Museum Photo)

A BQM-34A carrying a 500 lb bomb. (Vadnais Photo)

Air-Launched Aircraft Techniques

For one reason or another, the airborne launch platform idea has, so far, not proven feasible. The name which has been applied to this releasing of aircraft at altitude is called a "Parasite."

This concept has been manifested in two different ways. Either a parasite aircraft is carried outside the launching aircraft, or, as with other more conceptual concepts, the planes were carried within a mother ship. But it should be noted that in the earliest applications of this concept, the mother ship was a balloon! Several World War I concepts were tested, including a British effort with the deHavilland Hummingbird.

The first United States incursion into this strange game evolved around the Army's tiny Sperry Messenger bi-wing aircraft. The concept was born at McCook Field in 1922 using a balloon as the launch platform. The experimental work was carried out by Lawrence Sperry, who volunteered to test the feasibility of hooking on to the airship.

Keeping the propeller out of the airship's dangling trapeze grappling system proved to be the major operational problem during the tests. The little Messenger actually joined the airship by a hook that was mounted on its top wing. The hook's supporting structure was carried foward to form a unique propeller guard.

Shock absorbers were used to ease the jolt of contact. The Messenger's hook was designed to open when a 2 mph differential was experienced.

The first Messenger hook-up attempts took place on September 18, 1923, at the Army Airship Base at Scott Field, Illinois. On the first two passes of the initial test, the 850 pound aircraft missed contact, and then broke its propeller on the third. The tests were conducted directly over the field with just such an eventuality in mind.

But in September of 1924, success was finally realized with the first solid hook-up being accomplished. The airship was moving at a blazing 62 miles per hour when the hook-up was made, and

A twilight view of a Firebee missile under the wing of a C-130 transport. (Air Force Museum Photo)

C-130 mounting a pair of Firebee drones. (Air Force Museum Photo)

Sperry Messenger. (Air Force Museum Photo)

Hook-up mechanism for Sperry Messenger. (Air Force Museum Photo)

the Messenger realized its first air launch when it was released at 51 miles per hour.

The promising Messenger concept, though, proved to be very short-lived, with all the tiny aircraft becoming surplus in 1926. Less than 50 of the models were built, and the only surviving example of the model now rests in the Air Force Museum at Wright Patterson AFB, where it dangles in repose from the ceiling on a simulated trapeze.

But the Messenger was probably not the most famous of the U.S. balloon-mounted Parasites. That honor would undoubtedly belong to the Navy Sparrowhawk project. Officially designated the XF9C-2, the Sparrowhawk bi-wingers enjoyed great publicity in their association with two famous Navy balloons, the USS *Macon* and the USS *Akron*.

The hook-up configuration of the Sparrowhawk was not at all unlike the Messenger rigging. The program was carried out during the early 1930s, but the USS *Macon* went down with three of the attached Sparrowhawks. During the next few years, the remaining Hawks just faded into oblivion.

Forty years later, the balloon concept was reconsidered. One of the strangest concepts was studied in the early 1970s, the so-called "Megalifter," where a lighter-than-air vehicle would serve as an airborne launcher. Except in the Megalifter concept, the parasite never materialized. Other USAF studies concerned themselves with the pilotless strike and reconnaissance vehicles. Other ideas have also included airship-lofted ballistic missiles.

During the 1940s, the carrier vehicle switched from balloon platforms to aircraft, and a multitude of new ideas were examined.

For example, the British employed a concept as a means for getting a heavily-loaded seaplane airborne. The mother flying boat—the MAIA—carried a smaller seaplane, the Mercury, on a pylon.

The first separation flight was accomplished in 1938. The two aircraft made numerous flights with separations at altitude being accomplished. MAIA and Mercury flights were operating between Southhampton, England, and Alexandria, Egypt, when World War II broke out. Mercury was scrapped in 1941 after serving with a Dutch seaplane squadron attached to the RAF. MAIA was destroyed by a Luftwaffe bomb in May 1941.

The need for aircraft-borne fighters was realized early in World War II when B-17 bombers had to traverse a major power of the Axis bombing missions unescorted. P-51s and P-47s with drop tanks eventually solved the problem, but another concept, the Project MX-472—called the unconventional fighter—was started in 1942 to fill the need. The aircraft-carried fighter, though, would never reach beyond the paper design phase.

In the late 1940s, the French looked at the Parasite concept with the research of Rene Leduc, who developed his Leduc 0.10 et Monoplane which was mounted on pylons above its Languedoc 161 mother ship.

An early air-launched fighter program had this tiny Sperry Messenger carried aloft under an airship. (Air Force Museum Photo)

The concept of carrying lightweight fighters aloft was attempted by the U.S. Navy with its Sparrowhawk program and the Sparrowhawk air-launched fighter. (U.S. Navy Photo)

The Leduc 0.10 was first released as a glider at 3,000 meters altitude over Toulouse. The ramjet engine was then ignited at half-power and the plane reached a maximum speed of 450 miles per hour. A number of prototypes were constructed, some with ramjet power, while others used wingtip-mounted turbojet engines.

In a similar French program, a Sud-Ouest SO M1 glider was mounted on the same launching platform. The M1 was a scale test model for the purpose of investigating high-speed flight. It featured swept wings, a control system featuring small ailerons and spoilers, and a pilot emergency escape system.

The Germans during World War II were also involved in some interesting mutations of the piggy-back concept. The best-known concept involved the use of an unmanned Ju-88 twin-engined bomber loaded with explosives. Mounted on struts above, and attached to, the Ju-88 was an Me 109 fighter. The pilot of the fighter controlled the complete contraption. When near the target he cut the Ju-88 loose, and guided the explosive-laden bomber to its target. The innovative Germans also investigated the integration of an Me-109 and a DFS-230 Troop Glider.

One of the deadliest forms of parasiting was the World War II Japanese Kamakazi attacks using the air-launched Yokosuka Ohka. The Ohka was essentially a rocket-powered flying bomb with 2,646 pounds of high explosives on board.

This was a one-way mission for the pilot. The pilot was actually bolted into his cockpit sans parachute with the plane launched from atop a Mitsubishi G4M. It was necessary because the Ohka did not have enough power to lift off by itself.

After being released the pilot would start his engine, which had an operating time of about four minutes. These suicide weapons did not have an effect on the result of the war in the Pacific, but they did cause serious damage to many American ships.

The next attempt at parasiting occurred in the late 1940s when a so-called "wing-tip coupling" concept was introduced. The idea was brought to Wright Field after World War II by two German scientists.

Their unconventional theory implied that smaller aircraft could be hooked onto the wingtips of larger aircraft, thus creating an increased wing span and more lift from the more efficient wing. The scientists explained that the improvement would allow the small planes to be carried along with practically no drag penalty.

There was a small amount of USAF interest in the concept, and a minimal number of coupling experiments were actually carried out before interest was lost in the technique.

In the late 1940s, a weird aircraft air-launch program was undertaken with the XF-85 program. The bulbous little fighter was

Grappling hood on the XP-85. (Holder Photo)

XP-85 being deployed under wing of launching aircraft. (USAF Photo)

Hook on nose of FICON F-84. (Holder Photo)

built by McDonnell Douglas and was powered by a 3,000 pound thrust J-34 turbojet engine with enough fuel onboard for about 80 minutes of flight time.

The squatty Goblin sported eight control surfaces, two on the wings and six on the tail surfaces. Only two prototypes were built, with the initial air launch taking place in 1948 from an EB-29B mother ship.

The initial attempt at hooking up with the B-29 proved unsuccessful, as the Goblin fought the resistence of the air compression between the two aircraft. Later tests proved to be moderately successful. But the concept was quickly losing favor with the Air Force as mid-air refueling research showed greater promise.

Even a McDonnell proposal for a .9 Mach version of the craft with alternative delta and swept wings failed to generate any sizable interest. Shortly thereafter, the XF-85 program was terminated with only two hours 19 minutes of flight time between both aircraft.

Another Parasite program was carried out in the early 1950s under the acronym FICON, a program that combined the intercontinental range of the B-36 and the then-high performance of the F-84 Thunderjet.

The initial design of the concept looked extremely promising, so 25 F-84s were modified, to be called RF-84Ks. The planes were modified with a drooped horizontal stabilizer and a special hook-up mounting forward of the canopy. This arrangement allowed the planes to be retrieved or launched. The planes were carried in the bomb-bay of the B-36.

The RF-84Ks actually assumed an operational status at Larson Air Force Base, Washington, to become the lone FICON squadron in the Air Force. Regular operations using the modified planes commenced in early 1956, but almost immediately a series of hook-up problems caused cessation of the operations and the death of the concept. The special FICON equipment was removed, and the Thunderjets were converted back to their original F-84 configurations.

During the 1970s, the concept reared its head again with the so-called "Microfighter" concept. A Boeing proposal would have had the tiny planes carried by a 747 transport. The planes had a design weight of only 10,000 pounds. The use of lightweight composite materials was estimated to greatly reduce that weight.

Head-on view of F-84 FICON fighter. (Holder Photo)

F-84 FICON in launch position. (Air Force Museum Photo)

F-84 FICON fighter during testing. (Air Force Museum Photo)

The concept called for the 747 to carry up to ten of the fighters. The air-launch and recovery techique used a trapese system very similar to that used by the XF-85 program.

A wind tunnel program was carried out at the Air Force's Arnold Engineering Development Center. The tests concentrated on new shapes and configurations for the small air-launched, air-recoverable fighter. A number of different configurations were investigated before the program was shelved.

Air-Launched Missile Concepts

In addition to the many aircraft air-launched concepts, there were a number of paper studies in the 1970s and 1980s considering the possibility of the ultimate stand-off weapon—an air-launched ICBM. As radical as it seems, the concept was actually considered.

The concept involved the use of a Boeing 747-type transport and the solid-propellant three-stage Minuteman ICBM. The set-up had the missiles dropped from the plane and then ignited and stabilized once free of the launching aircraft.

The concept provided several advantages. First, there was the fact that the launching site was no longer a fixed ground location easily targeted. With the air-launch situation, the launching platform was constantly on the move, making location and targeting immensely difficult.

Then, too, theoretically, with the missile being fired from altitude, the energy expended for launch would no longer be required, and therefore would possibly increase the payload capability of the missile system.

Air-Launched Tactical Decoys

With the extremely large radar size of the B-52 bomber, there were efforts in the 1970s to disguise its size to detection. One of the most significant countermeasures carried for a number of years was the

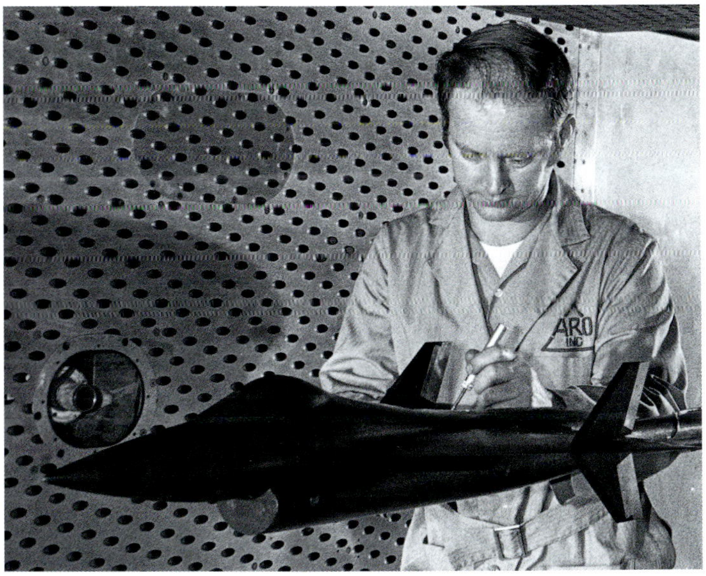

Testing occurred during the mid-1970s of a "Micro-Fighter," shown here during a wind tunnel test. (Air Force Museum Photo)

Inflight launch from a transport-type aircraft, the DC-10 in this case, was also attempted with the AGM-109. (USAF Photo)

This artist's concept shows a C-5 deploying proposed CMCA cruise missiles. (USAF Photo)

Artist's concept of B-52s being protected by GAM-72 Quail diversionary systems. (Air Force Museum Photo)

ADM-20 Quail decoy missile. Some 245 of the small diversionary missiles were ordered by the Air Force. The Quails were carried in the B-52 bomb bays, with engine ignition occurring after being dropped. The Quail was carried in so-called "Quick Load" clip-in packages in the B-52.

The little missiles effectively simulated a B-52 electronically to an enemy radar. The system degraded hostile air defense systems by its ability to simulate the flight and radar signature of the launching aircraft.

The Quail was powered by a General Electric J85 turbojet engine, and the missiles were guided by a pre-programmed autopilot. The system was integrated into the Strategic Air Command in 1961.

The size numbers for the system included a 13 foot fuselage length, five and one-half foot wing span, and 1,200 pound weight.

Another diversionary decoy program was given the name SCAD, for Subsonic Cruise Armed Decoy, which was to be the replacement for the Quail, but with one added dimension. The SCAD, in addition to its decoy mission, was also planned to carry a warhead. The SCAD was to be launched from the same launcher that carried the SRAM air-to-surface missiles.

The program was initiated in June 1972, and was canceled in early 1974. Another armed decoy missile was also considered, but it too was discarded.

Other Air-Launched Concepts

During the late 1990s, another interesting air launch concept was considered by the U.S. Air Force with its Eclipse Project. The program considered the use of a transport to tow a vehicle to altitude for an air launch. The technique was considered to be applicable for towing a large space vehicle to altitude where its rocket engine could be ignited. The concept was successfully demonstrated when a C-141 transport towed an F-106 fighter into the air.

An ADM-20 Quail is deployed from a B-52. (McDonnell-Douglas Photo)

This Quail is dwarfed by its B-52 carrier. (USAF Photo)

In the late 1990s, another variation of the air-launch technique was introduced with a number of proposed air-launched towed decoy programs.

Launch in the case of these decoys actually involved the decoys being deployed on a line from the launching aircraft and being towed behind the plane. Towed decoys have been tested on the F/A-18, B-IB, and other aircraft. The jury is still out on these new aircraft protection systems, but the early thinking is that they could be something very effective.

A Quail is mounted on its special ground support equipment. The Quail design was characterized by downturned wingtips. (McDonnell Douglas Photo)

Gallery

A pair of F-15 Eagles each carry defensive missile loads. (McDonnell Douglas Photo)

YF-22 fighter launches an AMRAAM during testing at the Pacific Test Center. (Lockheed Photo)

This 1991 photo shows an F/A-18 firing a Sidewinder during testing. (US Navy Photo)

An F-5E from the 57th TFW carries a pair of wingtip-mounted Sidewinder missiles. (USAF Photo)

The X-15 No. 2 rocket research aircraft hangs from the wing pylon mount of NASA's B-52 003 as it taxis to the runway for one of the 199 research flights in the X-15 program. The B-52 was one of two used to air launch the X-15s during the nine year program conducted by NASA Dryden Flight Research Facility, Edwards, CA, from 1959 to 1968. B-52 003, on loan to NASA from the Air Force, was the launch aircraft on 93 of the program's 199 flights. It was retired from service in 1969 and is now located at the Pima County Air Museum in Tuscon, AZ. (NASA Photo)

Bill Holder

Scott Vadnais

Air Launch is the third book that Scott Vadnais and Bill Holder have co-authored for Schiffer Publishing. The others are *Firepower* and *The C Planes.* Sott is a U.S. Air Force Major, serving as a Public Affairs Officer in a number of different assignments, including the Aeronautical Systems Center and the 325th Fighter Wing. Bill is a retired Aerospace Engineer, having worked for both Boeing and the Air Force. Besides his three books with Scott, he has done a number of other books for this publisher.

The pair hope to continue to collaborate on further aerospace book efforts in the future.